高等职业院校计算机类规划教材

MySQL 数据库系统原理
(Oracle Linux 8.3+MySQL 8.0.23)

李爱武　编著

北京邮电大学出版社
www.buptpress.com

内 容 简 介

本书包括关系型数据库基本理论以及 MySQL 数据库技术两部分。第 1～4 章为数据库基本理论部分，主要包括关系模型基本理论、ER 图及范式理论；第 5～20 章为 MySQL 数据库技术部分，主要包括在 Oracle Linux 上安装和运行 MySQL 服务器软件，SQL 语言，表空间，索引原理，MySQL 数据库体系结构，事务处理及锁，备份恢复，用户及权限管理等内容。

本书可作为高校计算机相关专业的数据库课程教材或参考书，也可供从事 MySQL 数据库相关技术工作的人员参考。

图书在版编目(CIP)数据

MySQL 数据库系统原理：Oracle Linux 8.3 ＋ MySQL 8.0.23 / 李爱武编著． -- 北京：北京邮电大学出版社，2021.7（2023.8重印）
ISBN 978-7-5635-6417-0

Ⅰ．①M… Ⅱ．①李… Ⅲ．①Linux 操作系统—高等学校—教材②关系数据库系统—高等学校—教材 Ⅳ．①TP316.85②TP311.138

中国版本图书馆 CIP 数据核字（2021）第 140167 号

策划编辑：马晓仟　　责任编辑：马晓仟　　封面设计：七星博纳

出版发行：北京邮电大学出版社
社　　　址：北京市海淀区西土城路 10 号
邮政编码：100876
发　行　部：电话：010-62282185　传真：010-62283578
E-mail：publish@bupt.edu.cn
经　　　销：各地新华书店
印　　　刷：唐山玺诚印务有限公司
开　　　本：787 mm×1 092 mm　1/16
印　　　张：18.25
字　　　数：478 千字
版　　　次：2021 年 7 月第 1 版
印　　　次：2023 年 8 月第 3 次印刷

ISBN 978-7-5635-6417-0　　　　　　　　　　　　　　　　　　　　　定价：49.00 元

・如有印装质量问题，请与北京邮电大学出版社发行部联系・

前　　言

　　数据库技术经过近60年的发展,关系型数据库逐渐一统天下。可以预见,关系型数据库在未来的几十年内还会居于主导地位。20世纪的最后20年,数据库市场主要被Oracle、微软、IBM三家公司控制。

　　进入21世纪后,Linux在服务器市场的份额越来越大,MySQL和PostgreSQL等开源数据库产品也迎来了黄金时代。考虑数据安全和成本等因素,国内越来越多的公司使用开源数据库产品作为服务器。几个大型科技公司,如阿里、腾讯、网易等,对MySQL的开源代码进行了二次定制开发,以满足企业自身的特殊要求。经过定制开发的MySQL甚至进入了一直由Oracle主导的金融行业,改变了人们对MySQL只部署于论坛、博客等非关键应用的印象。随着云计算技术的不断普及,以MySQL为代表的开源数据库将会迎来更广阔的市场。

　　本书包括关系型数据库基本理论和MySQL数据库技术两部分内容。在基本理论方面,本书力争用简短的篇幅和通俗的语言讲清其本质。在MySQL数据库技术方面,本书以简单的实验介绍相关知识、验证各种结论。在内容选取上,本书注重介绍数据库通用原理方面的知识,详细地讲解了SQL语言、索引、执行计划、统计信息、事务处理、锁、数据的导出与导入、重做日志等内容,为读者进一步学习MySQL和其他数据库产品知识打下基础。

　　本书的实验环境由Windows 10上的VMware 16.1.0虚拟机搭建,并在虚拟机上安装Oracle Linux 8.3和MySQL 8.0.23。SSH远程管理工具采用MobaXterm 20.5,MySQL图形管理工具采用MySQL Workbench 8.0.23。以上所有软件都可以安装在一台笔记本式计算机上,以方便模拟生产环境。与CentOS一样,Oracle Linux是Red Hat Enterprise Linux(RHEL)的"克隆",本书内容同样适用于RHEL和CentOS。

目　　录

第 1 章　数据库技术基础 ·· 1
1.1　数据库应用的场合 ·· 1
1.2　常用术语 ·· 1
1.3　数据库系统的构成 ·· 1
1.3.1　硬件 ·· 1
1.3.2　软件 ·· 2
1.3.3　人员 ·· 2

第 2 章　关系模型理论及主要产品 ·· 3
2.1　数据处理的历史 ·· 3
2.2　数据模型的概念 ·· 3
2.3　网状模型与层次模型 ·· 4
2.3.1　网状模型 ·· 4
2.3.2　层次模型 ·· 4
2.3.3　网状模型和层次模型的贡献及缺陷 ··· 5
2.4　关系模型的提出和成熟 ··· 5
2.4.1　关系模型要解决的问题 ·· 5
2.4.2　关系模型的提出与完善 ·· 6
2.4.3　IBM 的 System R 项目 ·· 6
2.4.4　加州大学伯克利分校的 Ingres 项目 ·· 6
2.5　关系模型的三个要素 ·· 7
2.5.1　关系模型的数据结构 ··· 7
2.5.2　关系模型的数据操作方式 ··· 8
2.5.3　关系模型中的完整性约束 ··· 8
2.5.4　关系型数据库的特点 ··· 8
2.6　主要关系型数据库产品介绍 ··· 9
2.6.1　传统商业数据库产品 ··· 9
2.6.2　开源数据库与 MySQL ·· 9

第 3 章 ER 模型 ... 10

3.1 数据库设计的主要步骤 ... 10
3.2 ER 模型的主要概念 ... 10
3.3 联系的映射约束 ... 11
3.4 ER 图转化为表 ... 11
3.5 使用 MySQL Workbench 进行数据库设计 ... 11

第 4 章 规范化理论 ... 17

4.1 引入范式理论的原因 ... 17
4.1.1 存在数据冗余的概念设计举例 ... 17
4.1.2 Insertion 异常 ... 18
4.1.3 Deletion 异常 ... 18
4.1.4 Update 异常 ... 18
4.2 第一范式 ... 18
4.3 第二范式 ... 19
4.4 第三范式 ... 20

第 5 章 安装和使用 MySQL ... 21

5.1 支持 MySQL 8.0 的操作系统 ... 21
5.2 在 VMware 16.1 虚拟机中安装 Oracle Linux 8.3 ... 21
5.2.1 下载 VMware 和 Oracle Linux ... 21
5.2.2 安装 Oracle Linux ... 22
5.3 使用 Windows SSH 工具管理 Oracle Linux 服务器 ... 24
5.3.1 使用 Windows 10 的 SSH 工具操作 Oracle Linux 服务器 ... 25
5.3.2 使用 Windows 10 的 scp 工具远程复制文件 ... 25
5.4 使用 MobaXterm 操作 Oracle Linux 服务器 ... 25
5.4.1 建立连接及基本 SSH 配置 ... 26
5.4.2 MobaXterm 基本配置 ... 26
5.4.3 在 Windows 和 Linux 之间拖动传输文件 ... 27
5.4.4 显示 X Window 图形 ... 27
5.5 使用 dnf 工具 ... 28
5.5.1 repo 配置文件 ... 28
5.5.2 管理 yum 源 ... 29
5.5.3 管理软件包 ... 30
5.5.4 dnf 模块管理 ... 30

5.6 安装与删除 MySQL 服务器软件 ·· 32
　　5.6.1 检查 MySQL 软件是否已安装 ·· 32
　　5.6.2 使用 Oracle Linux 系统附带软件包安装 ·· 32
　　5.6.3 管理 MySQL 服务 ·· 32
　　5.6.4 删除 MySQL 软件 ·· 33
　　5.6.5 使用 MySQL 官方 yum 源安装 ·· 34
5.7 升级 MySQL ··· 34
5.8 目录结构 ··· 35
5.9 连接至 mysqld 服务 ·· 35
　　5.9.1 连接至本地 mysqld 服务 ·· 35
　　5.9.2 连接至远端 mysqld 服务 ·· 35
5.10 mysql 工具执行常见操作 ·· 36
5.11 创建测试数据 ··· 37

第 6 章　SQL 查询语句 ··· 40

6.1 SQL 概述 ·· 40
　　6.1.1 SQL 语言的历史 ·· 40
　　6.1.2 SQL 的发音 ·· 41
　　6.1.3 SQL 查询的特点 ·· 41
　　6.1.4 SQL 标准 ··· 41
6.2 SQL 语言的主要类型 ·· 41
6.3 常用数据类型 ··· 42
　　6.3.1 数值类型 ·· 42
　　6.3.2 字符串类型 ··· 42
　　6.3.3 日期时间类型 ··· 42
6.4 简单的 SQL 查询语句 ·· 43
　　6.4.1 最简单的查询——只指定表 ··· 43
　　6.4.2 指定列 ·· 43
　　6.4.3 指定列别名 ··· 43
　　6.4.4 用 where 子句指定查询条件 ··· 44
　　6.4.5 使用 order by 子句给查询结果排序 ··· 44
　　6.4.6 分页查询 ·· 45
6.5 常用数值运算符及函数 ··· 45
　　6.5.1 数值运算符 ··· 45
　　6.5.2 常用数值函数 ··· 45
6.6 字符数据的处理 ·· 46

        6.6.1 字符串常量 ·· 46
        6.6.2 字符串模糊查询 ·· 46
        6.6.3 处理字符串中的特殊字符 ·· 47
        6.6.4 常用字符串函数 ·· 47
        6.6.5 利用正则表达式搜索字符串 ·· 51
    6.7 处理日期型数据 ·· 53
        6.7.1 获得当前日期时间 ·· 53
        6.7.2 日期型常量 ·· 53
        6.7.3 指定格式显示日期型列值 ·· 53
        6.7.4 抽取日期的指定部分 ··· 54
        6.7.5 获取时间差 ·· 54
    6.8 空值的处理 ··· 55
    6.9 分组汇总 ·· 56
        6.9.1 单独使用分组函数 ·· 56
        6.9.2 使用 group by 子句执行分组汇总 ··· 56
        6.9.3 having 子句 ··· 57
        6.9.4 order by 子句 ·· 57
        6.9.5 分组汇总查询小结 ·· 57
    6.10 子查询 ·· 58
        6.10.1 where 子句中使用子查询 ··· 58
        6.10.2 select 子句中使用子查询 ·· 59
        6.10.3 from 子句中使用子查询 ··· 59
        6.10.4 非相关子查询与相关子查询 ·· 59
        6.10.5 in 与 not in ·· 62
        6.10.6 exists 与 not exists ··· 64
    6.11 集合运算 ·· 65
    6.12 多表连接查询 ·· 65
        6.12.1 交叉连接 ·· 65
        6.12.2 内连接 ··· 66
        6.12.3 两种连接标准：SQL-86 与 SQL-92 ··· 67
        6.12.4 外连接 ··· 68
    6.13 构造复杂的查询语句 ·· 70
    6.14 SQL 查询的等效转换 ··· 72
        6.14.1 内连接与子查询 ·· 72
        6.14.2 in,exists,内连接 ·· 72
        6.14.3 not in,not exist,外连接 ·· 73

第 7 章 窗口函数 · · · · · · · · 74

7.1 over()子句构造窗口 · · · · · · · · 74
7.1.1 partition by 子句构造窗口 · · · · · · · · 74
7.1.2 order by 子句设置窗口内排序 · · · · · · · · 75
7.1.3 窗口内划分框架 · · · · · · · · 76
7.1.4 命名窗口 · · · · · · · · 79

7.2 窗口函数分类与示例 · · · · · · · · 79
7.2.1 汇总函数 · · · · · · · · 79
7.2.2 排名函数 · · · · · · · · 80

第 8 章 数据修改语句 · · · · · · · · 83

8.1 delete 语句 · · · · · · · · 83
8.1.1 简单的 delete 语句 · · · · · · · · 83
8.1.2 delete 语句使用 limit 子句 · · · · · · · · 83
8.1.3 delete 语句同时删除多个表的行 · · · · · · · · 83

8.2 update 语句 · · · · · · · · 84
8.2.1 简单的 update 语句 · · · · · · · · 84
8.2.2 update 语句修改多个表 · · · · · · · · 84

8.3 insert 语句 · · · · · · · · 85
8.4 replace 语句 · · · · · · · · 85

第 9 章 表及约束 · · · · · · · · 87

9.1 创建简单的表 · · · · · · · · 87
9.2 字符集及排序规则 · · · · · · · · 87
9.2.1 MySQL 与 Unicode 字符集 · · · · · · · · 87
9.2.2 排序规则与中文排序 · · · · · · · · 89
9.2.3 建表时设置字符集及排序规则 · · · · · · · · 89

9.3 建表时指定存储引擎 · · · · · · · · 90
9.4 使用 auto_increment 自增列 · · · · · · · · 90
9.5 自填充时间列 · · · · · · · · 91
9.6 约束 · · · · · · · · 92
9.6.1 约束的种类 · · · · · · · · 92
9.6.2 主键、非空及默认约束 · · · · · · · · 92
9.6.3 唯一、检查及外键约束 · · · · · · · · 93
9.6.4 对表增加约束 · · · · · · · · 95

>   9.6.5 删除约束 … 96
>   9.6.6 查询约束的信息 … 96
>   9.7 复制表 … 97
>   9.8 修改表的结构 … 97
>   9.8.1 修改列的数据类型 … 97
>   9.8.2 添加或删除列 … 98
>   9.8.3 修改列名 … 98
>   9.8.4 修改表名 … 98
>   9.8.5 清空表：truncate table … 98
>   9.8.6 删除表 … 99
>   9.9 查看表定义 … 99
>
> 第 10 章 分区表 … 100
>   10.1 分区类别 … 100
>   10.2 范围分区 … 100
>   10.2.1 单列范围分区 … 100
>   10.2.2 多列范围分区 … 102
>   10.2.3 增删范围分区 … 104
>   10.2.4 重组分区 … 105
>   10.3 列表分区 … 106
>   10.3.1 单列列表分区 … 106
>   10.3.2 多列列表分区 … 106
>   10.3.3 增删列表分区 … 108
>   10.3.4 重组列表分区 … 109
>   10.4 散列分区 … 109
>   10.4.1 普通散列分区 … 110
>   10.4.2 线性散列分区 … 110
>   10.4.3 键分区 … 111
>   10.4.4 重组散列分区 … 112
>   10.5 子分区 … 112
>   10.6 查询分区信息 … 113
>   10.7 改变表的分区类型 … 115
>   10.8 在 SQL 命令中直接操作分区 … 116
>
> 第 11 章 程序设计 … 117
>   11.1 用户变量 … 117

11.2 存储过程 118
11.2.1 存储过程的创建和执行 118
11.2.2 使用变量 118
11.2.3 使用 if 语句 119
11.2.4 使用 while 循环语句 120
11.2.5 使用输入参数及 SQL 语句 120
11.2.6 使用输出参数 121
11.3 函数 122
11.4 触发器 123
11.4.1 创建触发器的语法 123
11.4.2 old 和 new 的用法 124
11.4.3 模拟外键级联删除 125
11.4.4 约束检查 125
11.4.5 审计 127
11.4.6 查看触发器信息 128
11.5 查看可编程对象系统信息 128
11.5.1 使用 show 命令查看程序定义 129
11.5.2 使用 information_schema 系统库的 routines 和 triggers 视图 129
11.6 删除可编程对象 130

第 12 章 服务器体系结构 131

12.1 总体结构 131
12.2 内存结构 132
12.2.1 内存数据缓冲区 132
12.2.2 内存日志缓冲区 133
12.2.3 排序缓冲区和连接缓冲区 134
12.2.4 内部临时表内存 134
12.2.5 内存的自动设置 134
12.3 配置服务器和客户端参数 134
12.3.1 设置方式 135
12.3.2 使用命令行参数 135
12.3.3 使用配置文件 135
12.3.4 使用 set 命令 136
12.3.5 查看系统参数 136
12.4 事件日志文件 138
12.4.1 设置日志的时区和输出目标 138

		12.4.2　错误日志 ·· 139

		12.4.3　通用查询日志 ··· 140

		12.4.4　慢查询日志 ·· 141

	12.5　重做文件 ·· 141

		12.5.1　binary log ··· 141

		12.5.2　redo log ··· 145

	12.6　系统数据库 ··· 146

		12.6.1　mysql ··· 146

		12.6.2　sys ·· 147

		12.6.3　performance_schema ··· 150

		12.6.4　information_schema ··· 150

第 13 章　表空间和数据文件 ··· 151

	13.1　表空间的概念 ·· 151

		13.1.1　数据目录 ·· 151

		13.1.2　MySQL 的表空间分类 ·· 151

	13.2　system 表空间 ··· 152

		13.2.1　change buffer ·· 152

		13.2.2　doublewrite buffer ·· 152

		13.2.3　system 表空间的数据文件 ·· 152

	13.3　temporary 表空间 ··· 153

		13.3.1　会话临时表空间 ·· 153

		13.3.2　全局临时表空间 ·· 154

	13.4　undo 表空间 ·· 155

		13.4.1　默认 undo 表空间 ·· 155

		13.4.2　创建 undo 表空间 ·· 155

		13.4.3　查询 undo 表空间信息 ·· 155

		13.4.4　截断 undo 表空间 ·· 156

		13.4.5　删除 undo 表空间 ·· 157

	13.5　file-per-table 表空间 ··· 157

	13.6　general 表空间 ·· 159

	13.7　移动表所属的表空间 ··· 161

第 14 章　B 树索引 ··· 162

	14.1　B 树索引能把查询速度提高多少 ···································· 162

	14.2　一个使用索引的例子 ··· 163

14.3 主键索引的结构 165
14.4 普通索引的结构 168
14.5 索引能够提高查询速度的原因 172
14.6 需要创建索引的情况 172
14.7 如何知道一个查询是否使用了索引 172
14.8 不使用索引的情况 174
14.9 DML 语句对索引的影响 174
    14.9.1 insert 语句对索引的影响 174
    14.9.2 delete 语句对索引的影响 174
    14.9.3 update 语句对索引的影响 174
14.10 基于函数的索引 174
14.11 设置索引的可见性 175
14.12 多列索引 176
14.13 约束与索引 177
14.14 查询索引的系统信息 177
    14.14.1 show create table 177
    14.14.2 show index from 178
    14.14.3 查询 information_schema.innodb_indexes 178

第 15 章 执行计划 180

15.1 执行计划简单示例 180
15.2 执行计划的 select_type 属性 181
15.3 执行计划的 type 属性 186
15.4 执行计划的 ref 属性 189
15.5 执行计划的 Extra 属性 191

第 16 章 统计信息 195

16.1 统计信息的内容 195
16.2 统计信息的分类和收集 195
    16.2.1 永久统计信息 195
    16.2.2 临时统计信息 196
16.3 统计信息的存储和监控 197
    16.3.1 innodb_index_stats 系统表 197
    16.3.2 innodb_table_stats 系统表 198
    16.3.3 information_schema 中的统计信息视图 198
    16.3.4 show 命令 201

16.3.5 information_schema_stats_expiry 参数 ·················· 203

16.4 统计信息的更新 ·················· 203

    16.4.1 自动更新 ·················· 203

    16.4.2 手动更新 ·················· 204

16.5 列的直方图 ·················· 204

    16.5.1 直方图的作用 ·················· 205

    16.5.2 单值直方图和等高直方图 ·················· 205

    16.5.3 计算直方图 ·················· 205

    16.5.4 应用直方图实例 ·················· 207

第 17 章 事务处理 ·················· 210

17.1 事务的概念及应用实例 ·················· 210

    17.1.1 事务应用实例 1：银行转账 ·················· 210

    17.1.2 事务应用实例 2：超市收银 ·················· 210

17.2 事务的 ACID 属性 ·················· 211

    17.2.1 原子性 ·················· 211

    17.2.2 一致性 ·················· 211

    17.2.3 隔离性 ·················· 212

    17.2.4 持久性 ·················· 212

17.3 事务隔离级别 ·················· 212

    17.3.1 read uncommitted ·················· 212

    17.3.2 read committed ·················· 212

    17.3.3 repeatable read ·················· 213

    17.3.4 serializable ·················· 214

17.4 事务控制命令 ·················· 214

    17.4.1 commit 和 rollback 命令 ·················· 214

    17.4.2 设置事务模式 ·················· 215

    17.4.3 设置事务隔离级别 ·················· 215

    17.4.4 设置事务只读性 ·················· 216

17.5 并发控制要解决的问题 ·················· 217

    17.5.1 丢失更新 ·················· 217

    17.5.2 脏读 ·················· 218

    17.5.3 不可重复读 ·················· 219

第 18 章 锁 ·················· 220

18.1 MySQL 的锁类型和锁模式 ·················· 220

18.2 表锁 ··· 221
　18.2.1 表级 S 锁和表级 X 锁 ··· 221
　18.2.2 DDL 语句产生的元数据锁 ·· 222
　18.2.3 表级 IS 锁和表级 IX 锁 ·· 223
18.3 行锁 ··· 224
　18.3.1 S,REC_NOT_GAP 锁和 X,REC_NOT_GAP 锁 ································ 224
　18.3.2 S,GAP 锁和 X,GAP 锁 ·· 226
　18.3.3 next-key 锁 ·· 227
　18.3.4 insert intention 锁 ··· 230
18.4 insert 操作产生的隐式锁 ··· 233
18.5 外键对锁的影响 ·· 234
18.6 不同隔离级别下的加锁方式 ··· 235
　18.6.1 read uncommitted 与 read committed 隔离级别下的锁 ······················· 235
　18.6.2 repeatable read 隔离级别下的锁 ·· 240
　18.6.3 serializable 隔离级别下的锁 ··· 241
18.7 死锁 ··· 243

第 19 章 备份恢复 ·· 246

19.1 备份种类 ··· 246
　19.1.1 热备份与冷备份 ·· 246
　19.1.2 逻辑备份与物理备份 ··· 246
　19.1.3 全备份与增量备份 ·· 246
19.2 恢复种类 ··· 246
19.3 备份恢复工具 ··· 247
19.4 导出导入数据行 ·· 248
　19.4.1 一个简单示例 ··· 248
　19.4.2 使用 select into outfile 及 load data 执行导出导入 ······························ 249
　19.4.3 使用 mysqlimport 导入 ·· 249
　19.4.4 使用 mysqldump 导出 ·· 250
19.5 mysqlpump 执行逻辑备份 ·· 250
　19.5.1 mysqlpump 执行导出 ·· 250
　19.5.2 mysqlpump 的常用选项 ··· 251
19.6 使用 mysqlbinlog 导出二进制日志文件 ··· 252
19.7 基于时点的恢复 ·· 253

第 20 章 用户和权限管理 ·· 255

20.1 MySQL 用户的特点 ··· 255

- 20.2 预置用户 ... 255
- 20.3 一个关于用户及权限的简单示例 ... 256
- 20.4 用户管理 ... 256
  - 20.4.1 MySQL 的用户名标识 ... 257
  - 20.4.2 创建和删除用户 ... 257
- 20.5 密码管理 ... 257
  - 20.5.1 密码加密算法 ... 257
  - 20.5.2 设置密码过期和重用规则 ... 258
  - 20.5.3 设置密码试错次数上限及锁定天数 ... 258
  - 20.5.4 查询用户的密码相关属性 ... 259
  - 20.5.5 设置密码相关系统参数 ... 259
  - 20.5.6 设置密码复杂度规则 ... 260
- 20.6 对用户设置资源限制 ... 261
- 20.7 修改用户属性 ... 261
- 20.8 权限管理 ... 262
  - 20.8.1 MySQL 中的权限层次和名称 ... 262
  - 20.8.2 赋予和撤销权限 ... 262
  - 20.8.3 全局权限的部分撤销 ... 264
  - 20.8.4 查询用户的权限信息 ... 266
- 20.9 角色 ... 270
  - 20.9.1 创建角色并对其赋权 ... 270
  - 20.9.2 角色的赋予及撤销 ... 271
  - 20.9.3 激活角色 ... 272
  - 20.9.4 用户的默认角色 ... 273
  - 20.9.5 设置公共角色 ... 274
  - 20.9.6 角色相关信息查询 ... 274

参考文献 ... 276

第 1 章　数据库技术基础

随着计算机技术的快速发展,数据库应用的范围越来越广泛,日常生活对数据库应用的依赖越来越紧密。本章对数据库技术的相关背景和概念做简单介绍。

本章主要内容包括:
- 数据库应用的场合;
- 常用术语;
- 数据库系统的构成。

1.1　数据库应用的场合

随着移动互联网技术的飞速发展,人们可以使用网上银行、手机银行进行各种金融活动,使用淘宝、京东等网站或手机客户端软件购物,使用美团、饿了么点外卖,使用微信、QQ 等即时通信软件与朋友交流,使用嘀嘀打车、共享单车等出行。以上这些系统都离不开数据库的支撑,近年兴起的大数据技术更离不开数据库的支持。就业市场对数据库技术人员,如数据库管理、应用开发等方面人才的需求越来越大。

1.2　常用术语

数据库(Database,DB)是长期存储在计算机内的、有组织、有结构、可共享的数据集合。数据按一定数据模型组织和存储,具有较小冗余度、较高数据独立性和易扩展性。

数据库管理系统(Database Management System,DBMS)是位于用户和操作系统之间的一类系统软件。它用于科学、有效地组织和存储数据,高效地获取和维护数据。Oracle、DB2、SQL Server、MySQL、PostgreSQL 都是典型的 DBMS 产品。

数据库系统(Database System,DBS)是引入了数据库的计算机应用系统。

1.3　数据库系统的构成

数据库系统一般由硬件、软件及各种人员构成,下面分别介绍。

1.3.1　硬件

数据库系统的数据量和访问量一般很大,而且要保持常年 24 小时运转,整个数据库系统对硬件资源的要求很高。能够用作数据库服务器的计算机一般由 Oracle、IBM、HP 等著名计算机厂商生产,使用专用的 CPU、内存,数据存储采用磁盘阵列,性能和价格远远高于 PC 上的各种硬件。昂贵的服务器硬件限制了数据库应用系统的使用范围。

1.3.2 软件

数据库系统的软件主要包括操作系统、DBMS、前端应用软件。

大型数据库服务器一般采用专用服务器，安装专用的 Unix 操作系统，如 Oracle 的 SPARC 系列服务器产品只能安装 Solaris，而 IBM 公司的高端服务器只能安装 AIX。中小型应用对服务器硬件的要求稍低，一般使用 PC 服务器，安装 Windows 或 Linux 操作系统。

应用不同，DBMS 也不同。单用户应用使用桌面数据库产品就够了，如微软公司的 Access 产品，而大型、关键应用（指涉及金融或财务的应用）一般采用大型数据库产品，如 Oracle、DB2 或 SQL Server。

前端应用软件一般由另外的软件公司根据企业的具体要求进行定制开发。

随着开源软件和云计算技术的发展，越来越多的应用采用 Linux 系统或云架构运行其数据库服务器，这大大降低了软硬件购置和维护费用。典型的云平台包括亚马逊 AWS、微软 Azure 以及阿里云。当前很多数据库服务器由云上运行的 Linux+MySQL 等开源软件搭建。

1.3.3 人员

数据库系统涉及的人员主要包括系统分析和数据库设计人员、应用程序开发人员，数据库管理人员和终端用户。

系统分析人员负责应用系统的需求分析和规范说明，与用户及数据库管理人员讨论，确定系统的软硬件配置，并参与数据库系统的概要设计。系统分析人员要具备全面的软件工程知识及丰富的系统设计、开发经验。

数据库设计人员负责数据库中数据的确定，参加用户需求调查和系统分析，根据分析结果设计出应用所需的表结构。

应用程序开发人员负责设计和编写应用系统的程序模块及前端界面，需要掌握 SQL 语言及存储过程编写方法，并要掌握所使用的开发语言。

数据库管理人员负责管理和维护数据库系统，具体职责包括以下几个方面：决定数据库的存储结构和存取策略；确定数据库的安全性要求，完成用户和权限控制；负责数据库的备份和恢复；监控数据库运行，如空间使用情况，以及确定操作低效的成因并解决。

以上人员都要精通数据库理论及 DBMS 产品的使用，另外，数据库管理人员还要熟悉相关 DBMS 和操作系统的原理及操作，并具备一定的程序开发知识。

终端用户指数据库应用系统要服务的对象，这些人一般不需要具备计算机和数据库技术知识，例如使用银行自动取款机取款的人，就是银行数据库系统的终端用户；使用淘宝网购物的人就是淘宝数据库系统的终端用户。

第 2 章 关系模型理论及主要产品

关系型数据库是当前应用最广泛的数据库类型，其理论基础是关系数据模型。本章介绍关系型数据库和关系数据模型的基本概念。

本章主要内容包括：
- 数据处理的历史；
- 数据模型的概念及网状模型与层次模型；
- 关系模型的提出和成熟；
- 主要关系型数据库产品介绍。

2.1 数据处理的历史

从计算机工业早期开始，存储和处理数据一直是应用的核心。数据库技术是计算机科学中发展最快、应用最广的技术之一，已成为计算机应用系统的核心技术和重要基础。

计算机在 20 世纪 40 年代诞生之初主要用于科学与工程计算，只能处理数字，不能处理字母和符号，也缺乏数据处理所需的大容量存储器。

20 世纪 50 年代初，字符发生器（Character Generator）的发明，使计算机能显示、存储与处理字母及各种符号。此后，人们又成功将高速磁带机用于计算机存储器，这是对计算机得以进入数据处理领域，具有决定意义的两大技术进展。但是磁带速度慢，只能顺序读写，不是理想的存储设备。

1956 年，IBM 公司和 Remington Rand 公司先后实验成功磁盘存储器，它不但存取速度快、容量大，还可以随机读写，计算机数据处理便日益发展起来。

初期的数据处理软件只能进行文件管理，数据文件和应用程序一一对应，容易造成数据冗余、数据不一致和数据依赖问题。所谓数据依赖是指应用程序与数据的存储路径、存取方式密切相关，这种状况给程序的编制和维护造成很大困扰。后来出现了文件管理系统作为应用程序和数据文件之间的接口，在一定程度上提高了数据处理的灵活性。但这种方式仍以分散、互相独立的数据文件为基础，存在数据冗余、不一致、处理效率低等问题。

针对上述问题，各国学者、计算机公司纷纷开展研究，以求突破文件系统分散管理的弱点，实现对数据的集中控制，统一管理。结果出现了一种全新的高效数据管理技术——数据库技术。数据库技术的理论基础是数据模型，下面先介绍数据模型的概念。

2.2 数据模型的概念

数据模型的概念和系统论述由 Codd 在 1980 年首先提出。

数据模型是对现实世界数据特征的模拟和抽象。与针对某种具体应用的数据模型（如经济增长模型)不同,数据库领域的数据模型具备通用性。

数据模型包括3个要素。

(1) 数据结构

数据结构是对数据抽象后,呈现给用户的逻辑形式,与数据在磁盘上的物理存储形式无关,如关系模型的数据结构是表格。数据结构是数据模型的最重要属性,不同的数据模型具有不同的数据结构。

(2) 数据操作

数据操作是对数据模型中的对象可执行操作的集合,主要用于检索数据。

(3) 约束条件

约束条件是一组完整性规则,用来保证数据的正确性、有效性、相容性。

有3种数据模型对数据库的发展产生了重要影响:网状模型、层次模型及关系模型。

2.3 网状模型与层次模型

虽然网状模型和层次模型对数据库技术产生了很大影响,但因其固有的缺陷,已退出历史舞台,当前数据库市场上主要是关系模型数据库产品。下面对网状模型和层次模型的发展过程、贡献和缺陷做简单介绍。

2.3.1 网状模型

1964 年,美国通用电气公司的 Charles Bachman 设计了最早的、具备通用功能的 DBMS,名为集成数据存储(Integrated Data Store, IDS)。IDS 的数据结构采用了网状数据模型(Network Data Model),简称网状模型。

网状模型在 1969 年被数据系统语言会议(Conference on Data Systems Languages, CODASYL)联盟下面的数据库任务组(Database Task Group, DBTG)进行了规范化,数据库理论中的三级模式以及数据定义语言(Data Definition Language, DDL)、数据操纵语言(Data Manipulation Language, DML)即由此规范首次提出。

因为对数据库技术的贡献,Bachman 在 1973 年获得计算机科学的最高奖——图灵奖,他是第一个因为对数据库技术的贡献而获此殊荣的学者。

2.3.2 层次模型

1965 年,为了管理阿波罗登月计划中的零部件,Rockwell 公司与 IBM 公司合作开发了一套名为 ICS/DL/I 的计算机系统(Information Control System and Data Language/Interface)。1967 年,IBM 团队发布了 ICS 的第一个版本。1969 年,ICS 更名为 IMS/360(Information Management System/360)。IMS 采用了层次数据模型(Hierarchical Data Model),简称层次模型。

IMS 引入了程序和数据分离的想法,应用程序通过调用 DL/I 语言访问数据。程序和数据的分离开创了编写数据库应用程序的新模式,使得应用程序可以专注于数据操作,而不用去考虑复杂的磁盘数据访问方式,在数据独立性方面,层次模型比网状模型有了很大提高。

IMS 现在依然是 IBM 的一个重要产品,在数据库市场上还拥有一定份额。

2.3.3　网状模型和层次模型的贡献及缺陷

IDS 是第一个具备通用功能的数据库产品,开创了数据处理的数据库时代。在随后的 CODASYL 规范中提出了数据库的三层模式、DML 与 DDL 语言,这些概念一直沿用至今。

IMS 第一个提出了程序与数据分离的思想,使得数据独立性有了很大提高。

与后来的关系模型相比,网状模型和层次模型主要有下面几个缺陷。

（1）数据结构复杂

网状模型的数据之间相互连接成网状,层次模型采用树形结构存储数据,这些复杂结构使得添加、删除数据等操作的复杂性大大增加。

（2）数据独立性存在一定问题

两种模型中,数据之间用指针相连,需要程序员编写程序处理查询请求,数据处理算法会嵌入到应用程序中。为了顾及访问效率,算法的设计要参考编写程序时的数据状态,即应用程序与数据状态是相关的,从而存在一定程度的数据不独立,若数据状态发生改变,之前算法可能不再适用,需要不断修改应用程序。

应用程序和数据之间的依赖会导致功能扩充和维护方面的诸多问题。关系模型的提出正是为了解决这些问题。

2.4　关系模型的提出和成熟

以 IDS 和 IMS 为代表的数据库产品在 20 世纪 60 年代后期推动了数据库技术的发展,显著提高了数据处理效率,但经过几年的使用,其数据结构复杂、数据独立性差的缺点也逐渐暴露出来。出于解决这些问题的需要,在 20 世纪 70 年代初期,关系模型应运而生,并在随后几年逐步成熟起来。

此后的几十年,关系模型理论经受住了实践的考验,关系型数据库产品大量涌现,现在的数据库市场依然是关系型数据库的市场。

在关系模型的发展过程中,起决定作用的是两个事件:
- Codd 提出并完善了关系模型理论体系;
- IBM 的 System R 项目解决了关系模型在实际应用中的关键技术。

2.4.1　关系模型要解决的问题

IDS 和 IMS 数据库系统的数据独立性比较差,当数据库的逻辑结构或物理结构发生改变时,程序员要将大量时间用在应用程序的修改和维护上。为了改变这种状况,IBM 公司的 Codd 致力于构造一种能够更好地提供数据独立性的数据模型。

Codd 的建议包括以下三个方面:
- 用一种简单的结构存储数据;
- 用一种更高层次的、面向集合的语言访问数据;
- 不需要说明数据的物理存储方式。

这三个方面的实质都是数据独立性。用简单的结构存储数据是指数据的逻辑存储形式,简单的数据结构可以提供更好的逻辑数据独立性,从而在改变数据的逻辑结构时,应用程序不需要很大修改。使用更高层次的操作语言,不需要指明数据的物理存储位置,可以提供更好的数据物理独立性。

2.4.2　关系模型的提出与完善

1970 年，Codd 在其论文"A Relational Model of Data for Large Shared Data Banks"中提出了关系数据模型，同时也提出了在数据库设计中广泛使用的规范化理论。这篇论文被认为是数据库技术发展史上具有划时代意义的里程碑。

在此之后，Codd 继续致力于完善与发展关系数据模型理论，使数据库成为一门科学。因其对数据库技术的突出贡献，他获得了 1981 年度的图灵奖。

关系模型提出后，一大批关系数据库系统产品被开发出来并迅速占领市场，基于层次模型和网状模型的数据库产品很快走向衰败甚至消亡。

在关系模型理论转化为产品并推向实际应用的过程中，有两个研究项目起了关键作用，一是 IBM 的 System R 项目，二是加州大学伯克利分校的 Ingres 项目。

下面对这两个项目分别进行介绍，并介绍其各自衍生的数据库产品。

2.4.3　IBM 的 System R 项目

Codd 提出关系模型后，IBM 在 1973 年启动了由 W. F. King 领导的 System R 项目，R 表示 Relation，其目的是检验关系模型的诸多优点是否可以在实际系统中实现，特别是关系型系统能否像人一样，能够根据不同的数据环境状态，构造最优的数据访问算法。

System R 的开发分为三个阶段。

第一阶段：1974—1975 年，SQL 语言的设计和开发。

此阶段使用了 R. Lorie 开发的 XRM 关系型数据访问算法，R. Lorie 是 IBM 坎布里奇科学中心的研究人员。XRM 方法是一个单用户访问方法，不支持并发和恢复功能。在 XRM 的上层，使用一个 PL/I 语言编写的解释器执行 SQL 命令。

第二阶段：1976—1977 年，全功能、多用户系统的设计开发。

当前大型数据库的所有关键技术，在此阶段几乎都被实现了，如视图、基于代价的优化器、B 树索引、事务处理与锁、重做日志等功能。System R 只用了 80 000 行代码，仅 2.2 MB。

第三阶段：1978—1979 年，System R 在实际应用中的评估。

整个评估分为两个部分，一是在 IBM 的圣何塞研究实验室，二是在 3 个客户机构以及 IBM 的几个内部机构。在试用过程中，System R 的易用性和强大功能得到了用户肯定。项目组根据用户的建议增加了一些功能，如 SQL 语言中的 EXISTS、LIKE 关键字。

项目主要参与者 James Gray 因对数据库和事务处理的贡献获得了 1998 年度图灵奖。

System R 项目后来演化为 IBM 的 SQL/Data 产品，1981 年推出商业版，这是 IBM 公司的第一个关系型数据库产品，第二个商业版关系型数据库产品是其在 1983 年推出的 DB2。

Oracle 数据库产品是其公司创始人受 System R 技术的启发而开发出来的，世界上第一个商业版关系型数据库产品是 1979 年发布的 Oracle 2.0。

2.4.4　加州大学伯克利分校的 Ingres 项目

1973 年，受 System R 项目启发，加州大学伯克利分校的 Michael Stonebraker 和 Eugene Wong 也开始了相似的研究项目，名为 Ingres(INteractive Graphics REtrieval System)。

Michael Stonebraker 和 Eugene Wong 在 1980 年成立了 RTI 公司（Relational

Technology, Inc.),开发 Ingres 系列数据库产品,后来公司名称改为 Ingres 公司。Ingres 并未跟随 IBM 使用 SQL 语言,而是采用 Quel 语言。1985 年之后,由于 IBM 的强大影响力,数据库市场对 SQL 语言的认可度越来越高,Ingres 决定由 Quel 语言转换至 SQL。经过多次并购,Ingres 产品现在属于 Actian 公司。

1984 年,Ingres 软件的主要开发者 Robert Epstein 在加州伯克利创立了 Sybase 公司(SYstem dataBASE)。1987 年 5 月,发布了 Sybase SQL Server 系统,这是第一个具备 C/S 结构的数据库产品。在 20 世纪 80 年代和 90 年代,Sybase 数据库产品的市场份额曾一度只落后于 Oracle,居于第 2 位。

1993 年,微软公司和 Sybase 公司签订协议,购买了使用 Sybase 源代码开发数据库产品的许可,但这仅限于 Windows 系统,此即后来的 SQL Server。

Ingres 项目衍生的另一个重要产品是当前流行的开源数据库产品 PostgreSQL。

2001 年,Michael Stonebraker 进入麻省理工学院任职,因为其对现代数据库系统的实践和基础理论的贡献获得 2014 年度图灵奖。

2.5 关系模型的三个要素

虽然关系模型的提出已超过 50 年,关系模型在各方面有了很大发展,但它的核心特征未发生变化,这也是理论与技术的一个显著区别:技术的发展日新月异,理论却有足够的稳定性。下面从构成数据模型的三个方面考察关系模型。

2.5.1 关系模型的数据结构

关系模型的数据结构是关系(relation)。

关系是一个数学概念,是由元组(tuple)构成的集合,而元组具有若干属性(attribute)。

直观上,我们可以把关系看作由行(row)和列(column)构成的表(table),行对应关系中的元组,列对应关系中的属性。在关系型数据库中,行也称记录(record),列也称字段(field)。

表 2-1 和表 2-2 是一个企业的员工表及部门表,分别称为 emp 表及 dept 表。

表 2-1 emp 表

empno	ename	job	sal	deptno
7369	SMITH	CLERK	800	20
7499	ALLEN	SALESMAN	1600	30
7521	WARD	SALESMAN	1250	30
7566	JONES	MANAGER	2975	20

表 2-2 dept 表

deptno	dname	loc
10	ACCOUNTING	NEW YORK
20	RESEARCH	DALLAS
30	SALES	CHICAGO

严格来说,关系模型中的关系与通常所说的表有下面一些关键区别。

关系是元组的集合,在数学中,集合中的元素不能重复,元素之间也没有前后顺序,相应地,关系中的元组不能重复,元组之间没有前后顺序;而表中的行可以重复,一般为了用户查看方便,也会按照某种顺序排列。

关系的属性没有从左到右的顺序,而表上的列一般有固定的顺序。

关系的属性不能为多值,而表中的列可以是多值的,如上面的 dept 表增加一个 phone(电话)列,其值可以有多个。

2.5.2 关系模型的数据操作方式

关系模型的数据操作方式称为关系代数(relational algebra),由一系列针对关系的运算构成。这些运算作用在一个或多个关系上,返回的结果还是关系。关系运算的输出结果可以作为另一个关系运算的输入,从而可以像算术运算一样把关系运算表达式嵌套使用。

在 Codd 提出的关系代数理论中,包括 5 种基本运算(为了方便,以下把关系称为表)。

选择:返回表中满足指定条件的行。

映射:返回表中若干指定列。

积:返回表的行由参与运算的两个表的行两两横向拼接而成。

并:返回两个表的并集。

差:从一个表中除去属于另一个表中的行。

2.5.3 关系模型中的完整性约束

在关系模型中,一个元组的某些属性比较特殊,一般称为键(Key),包括下面 3 种形式。

候选键:能唯一标识一个元组的最少属性构成的集合。候选键可以有多个。

主键:选择为唯一地标识元组的候选键称为主键。一个关系的主键只能有一个。

外键:一个关系中,属性值要匹配于另一个关系(或自身)中的候选键的一个或多个属性集合。如上面 emp 表的 deptno 字段可以作为其外键引用 dept 表的 deptno 字段。

完整性约束在一定程度上保证数据库中的数据是正确的,如上面 emp 表中的 sal 列表示员工工资,其值一般不能为负数,这种约束有很多,也各不相同。

下面两种约束的地位比其他约束重要,是所有关系型数据库都要具备的。

- 实体完整性:主键属性不允许空值也不能重复。
- 引用完整性:外键的值或者为空,或者匹配于其引用的键值。

2.5.4 关系型数据库的特点

关系型数据库具有以下特点。

(1) 严格的理论基础

关系型数据库建立在关系模型的数学理论基础上,先有理论再有产品,而网状模型及层次模型是在产品出现之后再总结出来的。

(2) 简单的逻辑结构

关系型数据库中,实体和实体间的联系都用关系表示。对数据的检索结果也是关系。关系即表,而表的结构简单、清晰,可以得到更高的数据逻辑独立性。

(3) 面向集合的操作方式

关系数据库中,操作语言是面向集合的,数据的存取路径对用户透明(即用户不用理会),

具有更高的数据物理独立性,简化了数据库应用程序的维护工作。

2.6 主要关系型数据库产品介绍

数据库工业经过 50 多年的发展,大型商业数据库市场主要被以下 3 个关系型数据库产品占据:Oracle 公司的 Oracle,IBM 公司的 DB2,微软公司的 SQL Server。开源数据库市场由 MySQL 和 PostgreSQL 主导。

2.6.1 传统商业数据库产品

收购 Sun 公司之前,Oracle 数据库一直是 Oracle 公司的核心产品。Oracle 公司是当今世界数据库工业的巨人,在市场占有率方面处于绝对主导地位。在中国各行业特别是电信和金融行业中有广泛应用。其特点是功能强大,但价格昂贵,体系结构复杂,入门较困难。

DB2 是 IBM 公司的产品,由 System R 进化而来,在中国主要应用于金融业。

SQL Server 是微软公司的数据库产品,其初期主要采用 Sybase 的技术,直到现在,SQL Server 的体系结构及操作方式与 Sybase 依然很相似。SQL Server 和其他几个数据库产品的最大区别是:SQL Server 不支持 Unix,只能安装在微软的 Windows 和 Linux 平台(2017 版本开始支持 Linux)。SQL Server 继承了 Windows 操作系统简单易用的特点,价格也稍低一些。

2.6.2 开源数据库与 MySQL

随着 Linux 性能及稳定性的不断提高,开源应用越来越多。在数据库产品方面,MySQL 和 PostgreSQL 是两个典型代表。

PostgreSQL 起源于加州大学伯克利分校的 Ingres 项目,第一个开源版在 1996 年 8 月发布,由一些志愿者负责。

MySQL 来自瑞典 TcX 公司的一个内部数据库应用 Unireg,在 20 世纪 80 年代开发。作者为芬兰人 Michael Widenius。1995 年,Michael Widenius 对 Unireg 增加了 SQL 接口,与另外二人成立了 MySQL AB 公司。

MySQL 的 SQL 处理模块与数据访问模块分离,允许在底层使用不同的存储引擎。

2002 年,MySQL 发布 4.0 版本,集成 InnoDB 引擎,完善了对事务处理的支持。

2005 年,InnoDB 存储引擎被 Oracle 公司收购。2008 年,Sun 公司以 10 亿美元收购了 MySQL。2010 年,Oracle 公司收购了 Sun 公司,MySQL 成为 Oracle 公司的产品。

2018 年,跳过版本 6 和 7,发布 MySQL 8.0 正式版(8.0.11),支持窗口函数和 InnoDB 集群等功能,默认字符集为 utf8mb4。

MySQL 性能高、成本低、可靠性好,已经成为最流行的开源数据库,被广泛应用于中小型网站。随着 MySQL 的不断成熟,它也逐渐用于更多大规模的网站和应用,如维基百科、Google 和 Facebook 等网站,国内的新浪微博和一些大型论坛也使用 MySQL 数据库。

因其开源性质,很多公司在 MySQL 代码的基础上进行定制开发,使其满足不同场景的特殊需要,阿里、腾讯、网易等中国科技公司的数据库产品,甚至可以满足金融系统应用的严苛要求,使得 MySQL 进入以往只有大型数据库才胜任的领域,以 MySQL 为代表的开源数据库产品的市场占有率不断提高。

MySQL 的商业版称为 MySQL Enterprise Edition,开源社区版称为 MySQL Community Server。一般提到 MySQL,主要指开源社区版,本书只讨论开源社区版 MySQL。

第 3 章　ER 模型

数据库设计是数据库应用开发的重要组成部分,构造 ER 图是需求分析完成后,进行数据库设计的开始,对后续程序开发过程有重要的影响。

本章主要内容包括:
- 数据库设计的主要步骤;
- ER 模型的主要概念;
- ER 图转化为表;
- 使用 MySQL Workbench 进行数据库设计。

3.1　数据库设计的主要步骤

应用开发的起始阶段是需求分析。需求分析的主要任务是通过对客户的调查,得到数据库应用要完成的功能及要保存的数据。

数据库设计是在需求分析的基础上依序进行的 3 个步骤。
- 概念设计:根据需求分析的结果抽象出实际应用中的实体及联系,画出 ER 图。
- 逻辑设计:把概念设计得到的 ER 图转化为表的结构,即得到建表语句。
- 物理设计:在 DBMS 上执行建表语句,也包括存储规划和创建索引等任务。

概念设计和逻辑设计独立于 DBMS 产品。逻辑设计得到的表还要验证是否符合相应范式。

本章将主要讨论概念设计和逻辑设计,下一章将讨论范式理论,存储空间和索引的知识见后面相关章节内容。

3.2　ER 模型的主要概念

数据库设计的一个困难问题是数据库设计者、开发者及最终用户常常以不同的视角看待相同的数据,解决这一问题就需要用一种非技术的方式交流对数据的理解,保证数据库的设计反映企业的需求,ER 模型就是为满足这种需要而设计的。对于数据库设计人员,ER 模型也提供了一个直观、具体的图形作为开始。

ER 表示 Entity Relationship,ER 模型即实体-联系模型,是一种概念设计的建模方法,由陈品山(Peter Pin-Shan Chen)在 1976 年提出。

现实世界由一系列的实体及其**联系**构成。实体和联系又由若干个**属性**限定。

实体属性中能唯一决定一个个体的一个或多个属性称为主键,如学生成绩表的学号,这里的主键即关系模型中的主键。在数据库设计和关系模型理论中,主键都具有重要地位。

ER 模型是一种自顶向下的设计方法,从识别实体以及实体间的联系开始,然后再标识实体及联系的属性。

以一个网上书店的数据库应用系统为例,其涉及的实体包括 Book(书籍)、Publisher(出版社)等实体,出版社与书籍的联系是"publish"。Book 实体有 ISBN(国际标准书号)、title(书名)、price(价格)等属性,而 Publisher 有 name(名称)、address(地址)、contact(联系人)等属性。

3.3 联系的映射约束

映射约束是两个实体集的实体之间的对应关系。映射约束的种类有以下 3 种。

(1) 一对一

如把出版社和联系人看作两个实体,若一个出版社只有一个联系人,且一个联系人只为一个出版社服务,则这两个实体之间的联系为一对一。

(2) 一对多

如员工和部门两个实体,若一个员工只属于一个部门,一个部门可以有多个员工,则这两个实体之间的联系为一对多。

(3) 多对多

如作者和书籍两个实体,一个作者可以编写多本书,一本书也可以由多个作者编写,这两个实体之间的联系为多对多。

3.4 ER 图转化为表

把 ER 图转化为表即进行数据库逻辑设计,一般遵循以下步骤。
① ER 图中的实体名称转化为表名。
② 实体的属性转化为表的字段。
③ 实体之间的联系根据不同对应关系转化为表。

很明显,在上面 3 个步骤中,主要是第 3 个步骤。

对于一对一联系,把两个实体及联系的属性合并为一个表,两个实体的主键都可以选作新表的主键。

对于一对多联系,把"一"的一方的主键及联系的属性合并到"多"的一方,并且此主键作为"多"的一方的外键指向"一"的一方的主键。

对于多对多联系,把联系转化为一个独立的表。此表由两个实体的主键及联系的属性构成,并且这两个实体的主键作为新表的外键各自指向由两个实体转化而成的两个表的主键。

3.5 使用 MySQL Workbench 进行数据库设计

ER 图有几种不同的画法,例如,陈品山最初提出的 Chen 方法,用矩形表示实体,用椭圆表示属性;流行的统一建模语言(Unified Modeling Language,UML)方法把矩形分为上下两部分,上面是实体名称,下面为实体属性。

当前有很多数据库设计软件,如 PowerDesigner、ERWin 等。Workbench 是 MySQL 官方提供的数据库管理工具,也提供了数据库设计功能,可以用来进行概念设计(画 ER 图),以及后续的逻辑设计和物理设计。MySQL Workbench 的安装文件可以在 MySQL 官方网站上免费下载。安装过程很简单，这里略过。

下面以 emp(人员)和 dept(部门)实体为例，说明使用 MySQL Workbench 8.0.23 进行数据库设计的方法。

启动 Workbench 后，其初始界面如图 3-1 所示。左侧竖向排列的 3 个按钮分别为建立数据库连接、数据库设计、数据迁移。

单击左侧第 2 个按钮，然后单击界面中的加号按钮，打开建模界面。

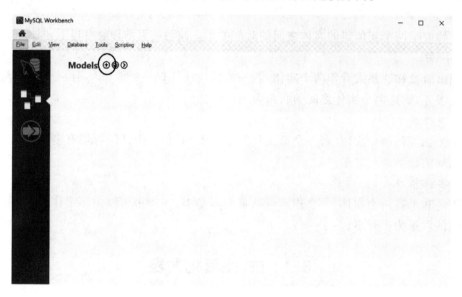

图 3-1　开始进行数据库设计

双击 Add Diagram 图标，打开 ER 图设计界面，其默认数据库名称为 mydb，如图 3-2 所示。

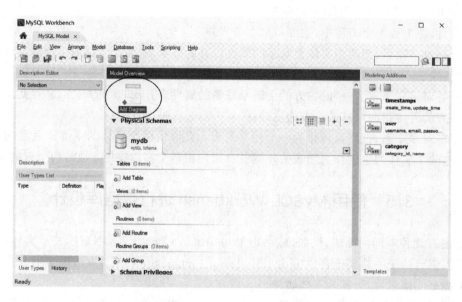

图 3-2　双击 Add Diagram 图标

在设计界面,主要使用左侧的表格按钮(即实体)和一对多关联按钮,如图 3-3 所示。

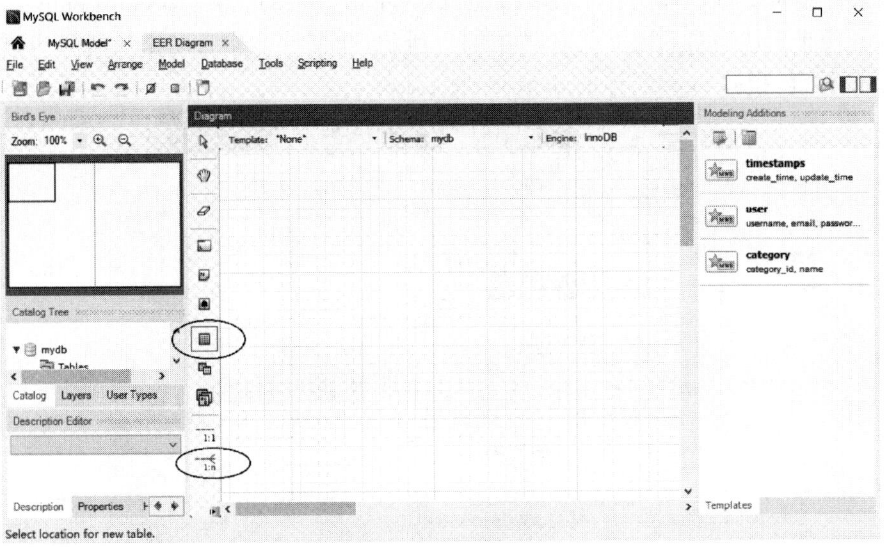

图 3-3　主要按钮

单击表格按钮后,再单击空白设计区域,即出现一个表格实体,重复以上操作,在设计区域创建两个表格实体 table1 和 table2,可拖动鼠标调整实体形状的大小。右击 table1 实体,选择"Edit 'table1'…",在界面下方会显示实体属性编辑窗口。编辑 Table Name,填入 emp,然后双击 Column Name 下面的空白区域,编辑属性名称,在 Datatype 部分选择合适的数据类型,如图 3-4 所示设置 4 个属性(empno、ename、job、sal),并选中右侧的 PK 多选框,把 empno 设置为主键。依照同样的步骤,编辑 dept 实体及其属性,如图 3-4 所示设置 3 个属性(deptno、dname、loc),把 deptno 设置为主键。

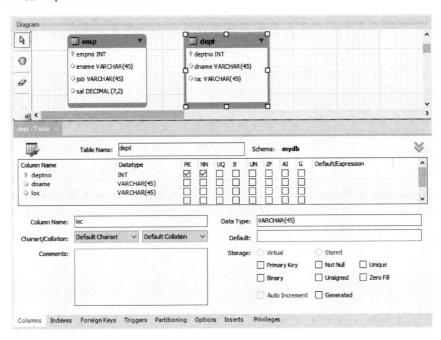

图 3-4　添加属性

dept 实体和 emp 实体的对应关系为一对多,下面建立两个实体间的关联。

关闭属性编辑窗口,单击"1:n"关联按钮,然后单击"多"的一方,即 emp,再单击"一"的一方,即 dept,两个实体会自动建立关联,如图 3-5 所示,并在 emp 中自动添加新列 dept_deptno,作为 emp 的外键,指向 dept 实体的 deptno。

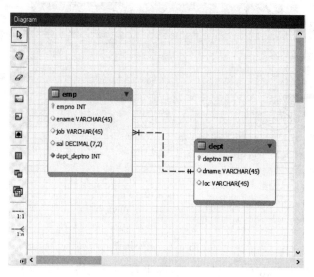

图 3-5 添加联系

单击"File"菜单,选择"Save Model"命令,保存以上 ER 模型,完成 ER 图设计。

继续执行逻辑设计,把以上 ER 图转化为建表语句。

在 MySQL 服务器创建用户 umodel,赋予其操作所有数据库的权限,关闭操作系统的防火墙,并确认服务器的 IP 地址(若尚未搭建实验环境,可先略过以下逻辑设计步骤)。

```
mysql> create user umodel identified by 'Umodel@1995';
Query OK, 0 rows affected (0.03 sec)

mysql> grant all on *.* to umodel;
Query OK, 0 rows affected (0.01 sec)

mysql> system systemctl stop firewalld
mysql> system hostname -I
192.168.199.130
```

单击"Database"菜单的"Forward Engineer …"命令,会弹出数据库连接设置窗口,如图 3-6 所示。

如图 3-7 所示,在连接选项设置窗口的 Hostname、Username,填入服务器 IP 地址及上面命令所创建的用户名称 umodel,然后单击"Store in Vault …",填入 umodel 用户的口令,单击"OK"按钮。

第3章 ER模型

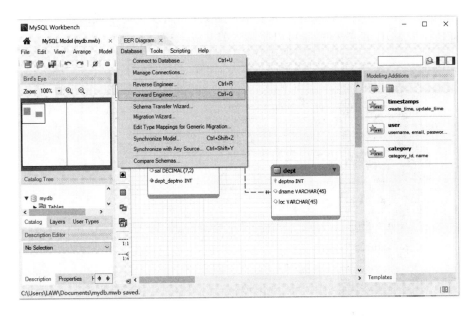

图 3-6　开始逻辑设计

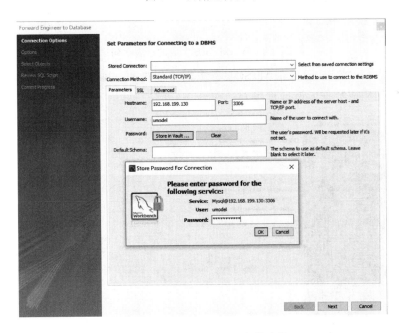

图 3-7　配置 MySQL 服务器连接

单击"Next"按钮,以下所有步骤均取默认设置,直到出现自动生成的建库及建表命令窗口,可以将其存为 SQL 脚本文件,也可以复制到剪贴板,如图 3-8 所示。

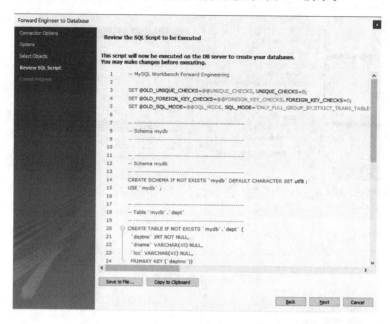

图 3-8　Workbench 生成的建表语句

继续单击"Next"按钮,即可在 MySQL 服务器中创建 mydb 数据库,以及 dept 和 emp 表,单击"Close"按钮完成数据库设计过程。

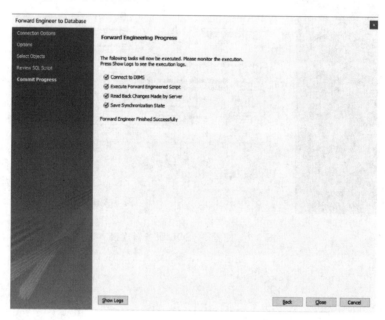

图 3-9　完成数据库设计

第4章 规范化理论

数据库规范化的目的是减少数据冗余,以避免数据异常。数据冗余即一份数据存储了多次,数据异常是由此引发的问题。规范化理论也称范式理论(Normal Form,NF)。范式理论应用于逻辑设计之后,用于确认表满足指定的范式等级(一般是第三范式)。

本章主要内容包括:
- 引入范式理论的原因;
- 第一范式、第二范式、第三范式的概念及转化方法。

4.1 引入范式理论的原因

在概念设计阶段,一个对象是抽象为单独的实体,还是作为另一个实体的属性,往往不容易确定。依据同样的需求分析,不同的数据库设计人员得到的表往往不同。有些设计结果可能存在数据冗余,即不满足范式理论。为了避免数据冗余,在逻辑设计后、物理设计前,需要应用范式理论验证是否每个表都符合指定的范式等级。

4.1.1 存在数据冗余的概念设计举例

如对一个公司的人力资源管理系统进行数据库设计时,在概念设计阶段抽象实体时,把部门属性当作员工属性,得到一个员工实体,转化为一个 emp 表,结构为

```
mysql> desc emp;
+----------+--------------+------+-----+---------+-------+
| Field    | Type         | Null | Key | Default | Extra |
+----------+--------------+------+-----+---------+-------+
| empno    | int          | NO   |     | NULL    |       |
| ename    | varchar(10)  | YES  |     | NULL    |       |
| job      | varchar(9)   | YES  |     | NULL    |       |
| mgr      | int          | YES  |     | NULL    |       |
| hiredate | date         | YES  |     | NULL    |       |
| sal      | decimal(7,2) | YES  |     | NULL    |       |
| comm     | decimal(7,2) | YES  |     | NULL    |       |
| deptno   | int          | YES  |     | NULL    |       |
| dname    | varchar(14)  | YES  |     | NULL    |       |
| loc      | varchar(13)  | YES  |     | NULL    |       |
+----------+--------------+------+-----+---------+-------+
10 rows in set (0.00 sec)
```

显然,此表的字段可以分为两部分,前一部分的7个字段描述员工个人信息,后一部分的3个字段描述员工所属部门信息。若 deptno = 10 的部门有 1 000 个员工,则这 1 000 条记录

的前 7 个字段的值是不同的,但后 3 个字段的值完全相同。

一个明显的问题是这样造成了空间浪费。除了空间浪费,还有下面几个更严重的问题。

4.1.2　Insertion 异常

新添加员工的部门数据要与现有同一部门的数据相同,否则就会造成表中的数据彼此不一致的情况。另外,如果一个新部门还没有员工,则此部门的信息不能添加到表中,表中也查不到这些部门的任何信息。

4.1.3　Deletion 异常

若一个部门的员工都辞职,其员工的记录都会被删除,虽然这个部门并未被撤销,但这个部门的信息在表中却不存在了。

4.1.4　Update 异常

如果一个部门有多个员工,而这个部门的名称发生了改变,则必须更新这个部门的所有记录的部门名称字段,如果漏掉一个,则会造成表中数据不一致的情况。这种异常更多是指修改纸版表格的情形。

减少数据冗余的方法是拆分表。拆分表时所要遵循的规则是范式理论要解决的问题。

4.2　第一范式

对表进行规范化处理的第一步是使其满足第一范式。

如果一个表没有多值字段,则这个表满足第一范式。要把包含多值字段的表转换为第一范式,可以将多值字段移出,与原表主键一起构成一个新表。

关系定义中,要求列为单值。多值字段虽然不会造成明显的数据冗余,但却给表的数据维护带来麻烦。

表 4-1 所示是一个网上书店系统用到的 Book 表及一行示例数据。ISBN 可以作为主键,显然 author(作者)字段是多值的。

表 4-1　不满足第一范式的 Book 表

ISBN	title	pubDate	price	publisher	author
7-04-007494-X	数据库系统概论	2002-1	25.10	高等教育	萨师煊,王珊

把多值字段 author 移出,和原表主键 ISBN 一起构成一个如图 4-2 所示的新表,其主键为 ISBN 和 author 列构成的复合主键。

表 4-2　ISBN_Author 表

ISBN	author
7-04-007494-X	萨师煊
7-04-007494-X	王珊

移出 author 字段后,原表变为如表 4-3 所示的新的 Book 表。

表 4-3　新的 Book 表

ISBN	title	pubDate	price	publisher
7-04-007494-X	数据库系统概论	2002-1	25.10	高等教育

经过这个过程,一个表分解为了两个表,这两个表都满足第一范式。

4.3　第二范式

如果一个表为第一范式,而且非主键字段完全依赖于主键字段,则此表满足第二范式。把非第二范式的表转化为第二范式,可以将存在部分依赖的非主键字段移出,与其所依赖的部分主键字段构成一个新表。

一个表不满足第二范式的前提条件是:其主键是多字段构成的复合主键,如果一个表的主键由单个字段构成,且没有多值字段,则必然满足第二范式。

表 4-4 中的主键为复合主键,由 ISBN 和 authorID(作者编号)两个列构成。

表 4-4　不满足第二范式的 Book 表

ISBN	title	pubDate	price	publisher	authorID	authorName	authorEmail

表的一部分非主键字段 title、pubDate、price 及 publisher 依赖于主键中的 ISBN 字段,而另一部分非主键字段 authorName、authorEmail 依赖于主键中的 authorID 字段。显然此表不满足第二范式。

把非主键字段 title、pubDate、price 及 publisher 移出原表,与其依赖的部分主键 ISBN 构成如表 4-5 所示的新表:

表 4-5　满足第二范式的 Book 表

ISBN	title	pubDate	price	publisher

把非主键字段 authorName、authorEmail 移出原表,与其依赖的部分主键 authorID 构成如表 4-6 所示的新表:

表 4-6　满足第二范式的 Author 表

authorID	authorName	authorEmail

原表未移出的字段构成如表 4-7 所示的 Book-Author 表:

表 4-7　满足第二范式的 Book_Author 表

ISBN	authorID

经过以上过程,原表被分解为 3 个表,均满足第二范式。

4.4 第三范式

如果一个表满足第二范式,并且没有非主键字段传递依赖于主键字段,则这个表满足第三范式。把非第三范式的表转化为第三范式,可以将存在传递依赖的字段从原表移出,与其直接依赖的字段构成一个新表。

表 4-8 所示是 emp(员工)表结构,主键为 empno(员工编号)。

表 4-8 不满足第三范式的 emp 表

empno	ename	job	salary	deptno	dname	loc

表中与部门相关的 dname(部门名称)和 loc(部门地址)字段直接依赖于 deptno(部门编号),而非直接依赖于主键 empno 字段,即 dname 字段传递依赖于主键 empno,因此此表不满足第三范式。

把存在传递依赖的非主键字段 dname 和 loc 从原表移出,与其直接依赖的 deptno 字段构成如表 4-9 所示的新表:

表 4-9 满足第三范式的 dept 表

deptno	dname	loc

原表中未移出的字段构成如表 4-10 所示的 emp 表:

表 4-10 满足第三范式的 emp 表

empno	ename	job	salary	deptno

经过以上过程,原表分解为两个表,均满足第三范式。

除了这 3 个常用范式,还有 BC 范式、第四、第五等范式。一般在实际应用中,满足第三范式就足够了。如果继续分解表,使其满足更高范式,虽然数据冗余更少,但一个查询可能会涉及更多的表。

第 5 章 安装和使用 MySQL

生产环境下，一般使用 Windows 下的 SSH(Secure Shell)客户端工具连接至 Linux 服务器进行远程操作。本书所述内容的实验环境为在 Windows 下的 VMware 虚拟机中运行 Linux 服务器，在 Linux 服务器上运行 MySQL 服务器软件，这样可以用一台机器构造一个与生产实际相似的实验环境。

本章主要内容包括：
- 在 VMware 虚拟机中安装 Oracle Linux；
- 使用 Windows 10 的 SSH 工具操作 Oracle Linux 服务器；
- 使用 SSH 工具 MobaXterm 操作 Oracle Linux 服务器；
- dnf 工具的使用方法；
- 使用 dnf 工具在 Oracle Linux 上管理 MySQL 软件的安装、删除和升级；
- 使用 systemctl 管理服务；
- 使用 mysql 客户端工具连接 MySQL 服务器；
- 使用 mysql 执行 SQL 语句和 SQL 脚本文件；
- MySQL 软件的目录结构简介。

5.1 支持 MySQL 8.0 的操作系统

MySQL 8.0 可安装在当前所有的主流操作系统上，如 Windows、Linux 及 macOS 等系统。

Red Hat Enterprise Linux 是使用最广泛的企业级 Linux 发行版，简称 RHEL，但其协议对个人用户不太友好。Oracle Linux 及 CentOS 都是 RHEL 的克隆，对个人用户使用基本不加限制。CentOS 已宣称在版本 8 之后不再支持 RHEL 稳定版，因此本书使用 Oracle Linux 搭建实验环境，所有内容均适用于 RHEL 或 CentOS。

5.2 在 VMware 16.1 虚拟机中安装 Oracle Linux 8.3

本书所述内容的实验环境由 Windows 10 上的 VMware 16.1.0 虚拟机搭建，并在虚拟机上安装 Oracle Linux 8.3 和 MySQL 8.0.23。SSH 远程管理工具采用 MobaXterm 20.5，MySQL 服务器管理工具采用 MySQL Workbench 8.0.23。

以上所有软件可以安装在一台笔记本式计算机上，方便模拟生产环境完成各种实验。

5.2.1 下载 VMware 和 Oracle Linux

VMware 有 30 天的试用期，下载 VMware 16.1.0 Windows 版本的安装文件(大小为 621 MB)，可

通过下面的网址：

https://download3.vmware.com/software/wkst/file/VMware-workstation-full-16.1.0-17198959.exe

在 Windows 上安装 VMware 很简单，这里略过。

Oracle Linux 8.3 的完整 ISO 安装文件（大小为 8.6 GB）下载网址为：

http://yum.oracle.com/ISOS/OracleLinux/OL8/u3/x86_64/OracleLinux-R8-U3-x86_64-dvd.iso

以上两个文件可直接下载，无须用户注册。

http://yum.oracle.com 是 Oracle 的开源软件库网站，其中的资源都可以免费下载。

5.2.2　安装 Oracle Linux

使用虚拟机包括创建虚拟机和安装操作系统，创建虚拟机主要是配置虚拟磁盘。以下是主要步骤。

VMware 启动后的界面如图 5-1 所示。

图 5-1　启动 VMware Workstation

单击"创建新的虚拟机"按钮，在新窗口中选择默认的"典型（推荐）"。若已有安装完毕的虚拟机，则可以单击"打开虚拟机"将其开启。

单击"下一步（N）"，在新窗口选择"安装程序光盘映像文件（iso）"，并单击"浏览"按钮，找到已下载的 Oracle Linux 8.3 的 ISO 安装映像文件。

单击"下一步（N）"，在新窗口填入虚拟机名称并选择安装目录，操作系统安装文件以及后续在操作系统执行的各种任务产生的数据都会保存在这个目录下。

单击"下一步（N）"，设置磁盘文件，各项均取默认。单击"下一步（N）"，出现总结窗口。单击"完成"开启虚拟机，开始安装过程。

鼠标点中开启后的虚拟机界面（若释放鼠标回主机，可同时按 Ctl+Alt 键），按向上方向键选择"Install Oracle Linux 8.3.0"，略过磁盘检查，按回车键开始安装操作系统，如图 5-2 所示。

图 5-2　开始安装过程

下面开始安装前的配置过程,第一个步骤是选择安装过程的语言,默认英文即可。

单击窗口右下角的"Continue"按钮,出现"INSTALLATION SUMMARY"界面,此界面用于系统配置及功能选择。所有属性分为 4 个部分,每部分按下面说明设置。

(1) LOCALIZATION

Time & Date:选择"Asia/Shanghai"时区,并开启"Network Time"(需先开启网络)。

(2) SOFTWARE

Software Selection:选择"Minimal Install",即最小安装。

(3) SYSTEM

Installation Destination:设置分区,默认即可(点开后,直接单击"Done"按钮)。

Network & Hostname:开启网络并设置机器名称。

(4) USER SETTINGS

单击"Root Password",设置超级用户 root 的口令。

图 5-3 所示是所有设置完成后的界面。

单击"Begin Installation"开始安装软件。安装完成后,单击"Reboot System"重启虚拟机 Linux 系统,如图 5-4 所示。重启完成后,鼠标点中虚拟机界面,输入登录账号 root 及其口令(口令不显示),然后执行 hostname -I 查看其 IP 地址以备用。

图 5-3 功能选择

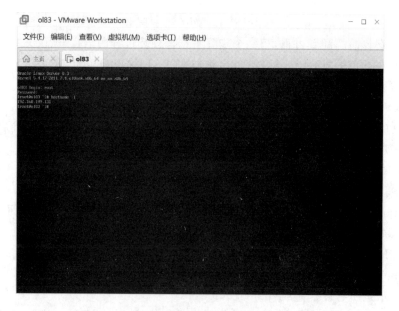

图 5-4 启动 Oracle Linux

5.3 使用 Windows SSH 工具管理 Oracle Linux 服务器

在生产环境下,系统管理员一般要远程管理 Linux 服务器。Windows 10 已附带客户端管理工具 SSH 和远程复制工具 scp。也可以选择功能更强大的第三方 SSH 工具,如 PuTTY、

Xshell,或本书使用的 MobaXterm 等软件。

5.3.1 使用 Windows 10 的 SSH 工具操作 Oracle Linux 服务器

假定 Linux 服务器的 IP 地址为 192.168.199.131,打开 Windows 系统的"命令提示符"工具,以 root 用户连接成功后,操作此服务器,如图 5-5 所示。

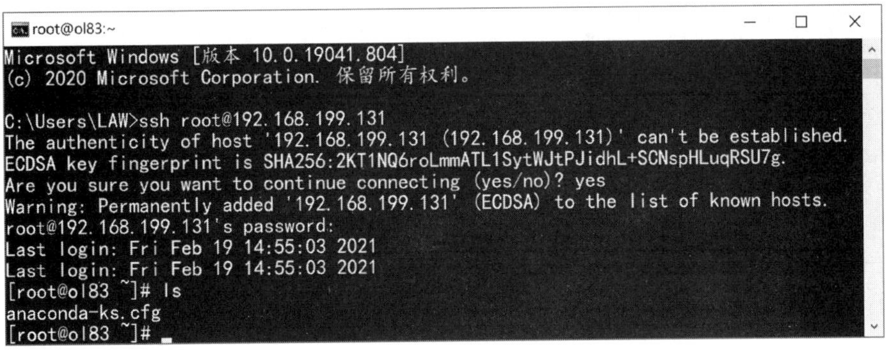

图 5-5　用 Windows 10 的 SSH 工具连接 Linux 服务器

5.3.2 使用 Windows 10 的 scp 工具远程复制文件

图 5-6 中的第 1 个命令,把 Windows 下载目录中的 scott.sql 文件复制至 Linux 的/root 目录。

图 5-6 中的第 2 个命令,把 Linux 的/root 目录下的 scott.sql 文件,复制至 Windows 的 D:/根目录,并重命名为 scott131.sql。若复制整个目录,可以使用 scp -r。

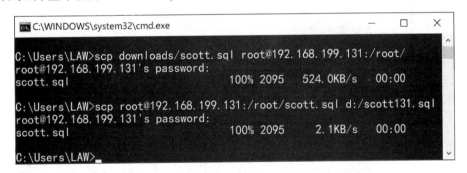

图 5-6　使用 Windows 10 的 scp 工具远程复制文件

5.4 使用 MobaXterm 操作 Oracle Linux 服务器

与其他 SSH 工具相比,MobaXterm 支持多个会话;它内嵌 X Server,可显示 X Window 图形;支持 SSH 文件传输协议(SSH File Transfer Protocol,SFTP),可以用鼠标拖动的方式在 Windows 和 Linux 之间传输文件。

MobaXterm 分为免费的家庭版和收费的专业版,家庭版功能即可以满足多数要求。家庭版又分为便携版(不需安装)和安装版。

5.4.1 建立连接及基本 SSH 配置

启动 MobaXterm 后,其初始界面如图 5-7 所示。

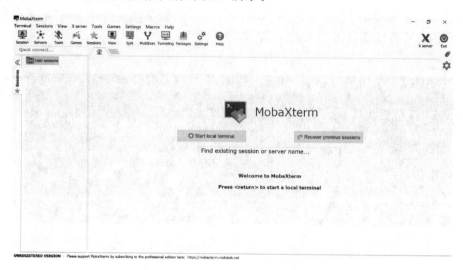

图 5-7　MobaXterm 的启动界面

在左侧栏右击"User sessions",选择"New session"创建服务器连接。在新窗口中单击"SSH"按钮,填入 IP 地址和登录账号(这里用 root)。单击"OK"按钮,建立连接,连接后,出现如图 5-8 所示的界面。

图 5-8　连接至服务器

5.4.2 MobaXterm 基本配置

(1) SSH 部分

取消勾选"Display SSH banner",连接成功时,不显示版本信息。

取消勾选"GSSAPI Kerberos",此项用于验证客户端机器的 IP 是否合法。

（2）Terminal 部分

勾选"Paste using right-click"，单击右键粘贴选中内容（选中的内容自动复制，不需设置）。

5.4.3 在 Windows 和 Linux 之间拖动传输文件

建立连接后，在左侧会出现新的 sftp 选项卡，是当前用户在 Linux 的主目录内容，单击向上的按钮，可以在目录间切换。可以用鼠标拖动的方式，在当前 Linux 目录和 Windows 目录之间传输文件。

5.4.4 显示 X Window 图形

为了节省资源，Linux 服务器一般不安装图形相关组件。MobaXterm 内置了 X Server，可以运行 X Window 图形程序，如 MySQL Workbench。只要设置 DISPLAY 环境变量通知服务器，把 X Window 发送至哪里。为了考查图形显示，先安装图形测试工具。

开启 yum 源 ol8_codeready_builder，然后安装图形测试软件包 xorg-x11-apps。

```
# dnf config-manager -- enable ol8_codeready_builder
# dnf -y install xorg-x11-apps
```

软件包中包含了 xclock、oclock、xeyes 以及 xlogo 等测试图形显示功能的小工具。

假定 MobaXterm 运行的 Windows 机器 IP 地址为 192.168.1.104，则可以执行下面命令运行 xclock 测试程序。

```
# export LC_ALL = C
# export DISPLAY = 192.168.1.104:0.0
# xclock
```

LC_ALL 环境变量用于设置字符集，以避免出现字符显示警告信息。DISPLAY 环境变量用于设置 X Window 发送至的机器地址。图 5-9 所示是实际效果。

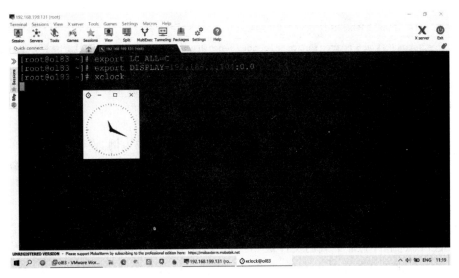

图 5-9　显示 X Window

5.5 使用 dnf 工具

yum 是 RHEL 系列操作系统的软件管理工具,以解决安装 rpm 包时的软件依赖问题,yum 工具使用一个 install 命令即可自动完成软件的下载和安装。

yum 在 RHEL 8 版本被替换为更先进的 dnf 工具,dnf 兼容 yum,仍然使用 yum 的各种配置文件。在 RHEL 8 系列系统中,yum 是 dnf-3(即 dnf)的软链接,执行 yum 即执行 dnf。本书使用 dnf 工具,命令中的 dnf 均可以替换为 yum 执行。

安装 MySQL 软件,dnf 是最简单的方式,本节简单介绍 dnf 的用法。

5.5.1 repo 配置文件

下载安装包时,dnf 工具要知晓下载地址。每个下载地址为一个软件库(repository,repo),也称 yum 源。这些位置信息存储在/etc/yum.repos.d 目录下的若干 repo 配置文件中,repo 文件名称不影响文件的下载和安装,但扩展名 repo 不能更改。

初次安装 Oracle Linux 8.3 后,默认有两个 repo 配置文件。

```
[root /etc/yum.repos.d 2021-02-20 22:08:58]
# ls
oracle-linux-ol8.repo   uek-ol8.repo
```

oracle-linux-ol8.repo 是与 RHEL 系统兼容的 repo 文件,uek(Unbreakable Enterprise Kernel)是 Oracle 创建的一个 Linux 内核,和此内核相关的软件库信息保存在 uek-ol8.repo 中。

repo 文件是文本文件,可手工修改。下面是 oracle-linux-ol8.repo 中的第 1 个 yum 源。

```
[root /etc/yum.repos.d 2021-02-21 08:31:01]
# cat oracle-linux-ol8.repo
[ol8_baseos_latest]
name = Oracle Linux 8 BaseOS Latest ($basearch)
baseurl = https://yum$ociregion.oracle.com/repo/OracleLinux/OL8/baseos/latest/$basearch/
gpgkey = file:///etc/pki/rpm-gpg/RPM-GPG-KEY-oracle
gpgcheck = 1
enabled = 1
...
```

每个 yum 源有 6 个属性,分为 6 行描述。以上面的 yum 源为例,说明一下每行的含义。

- [ol8_baseos_latest]:repo ID,dnf 以此 ID 管理 yum 源。此值不影响文件下载。
- name:repo 名称,可以看作对 yum 源的注释。此值不影响文件下载。
- baseurl:yum 源对应的网址,此项最重要,不能随意更改。
- gpgkey:使用此密钥验证下载的文件是否被非法更改。
- gpgcheck:是否验证下载的文件被非法更改,1 验证,0 不验证。
- enabled:此源是否开启,0 关闭,1 开启。查找文件时忽略关闭状态的 yum 源。

6 个部分中,前 3 项是必须的。若忽略后 3 项,enabled 默认为 1,gpgcheck 为 0。

用 dnf 下载安装软件时,需要把包含相关 yum 源的 repo 文件添加至/etc/yum.repos.d 目录,或者用文本编辑工具把 yum 源属性加入现有 repo 文件。

5.5.2 管理 yum 源

(1) 查看 yum 源

每个 repo 文件都配置了多个 yum 源,下面命令分别查看开启、关闭以及所有状态的 yum 源,enabled 是默认选项,可以省略。

```
# dnf repolist enabled
# dnf repolist disabled
# dnf repolist all
```

可以附加包含通配符的字符串作为筛选条件。

```
# dnf repolist enabled *base*
```

(2) 开启和关闭 yum 源

使用 dnf 的 config-manager 插件可以设置 yum 源的开启和关闭,以 repo ID 为操作标识。

```
# dnf config-manager --enable ol8_UEKR6
# dnf repconfig-manager --disable ol8_UEKR6
```

以上设置会存入相应 repo 配置文件。

也可以手工修改 repo 配置文件的 enabled 为 1 或 0,开启或关闭 yum 源。

(3) 增加新的 yum 源

增加新的 yum 源,一个方法是将其网址加进现有 repo 文件,或用网址创建一个新文件,repo ID 和 repo 名称可以自行指定。若存在对应的 rpm 文件,则可以执行 dnf install 命令,像安装软件包一样在/etc/yum.repos.d 目录下生成 repo 文件。

如把 Oracle 提供的 MySQL 软件 yum 源网址加入 oracle-linux-ol8.repo 文件,可执行下面的操作。

打开网址 http://yum.oracle.com/oracle-linux-8.html,此页面保存了 Oracle Linux 8 的各种 yum 源信息。找到 MySQL 8.0,打开 x86-64 的链接,复制页面右上角的 yum 源网址,加至/etc/yum.repos.d/oracle-linux-ol8.repo 文件(或其他 repo 文件)末尾,repo ID 和 repo name 自行指定如下。

```
[oracle_mysql8]
name = Oracle MySQL 8
baseurl = http://yum.oracle.com/repo/OracleLinux/OL8/MySQL80/community/x86_64
```

执行 dnf repolist 命令,确认新的 yum 源已经生效。

```
# dnf repolist *mysql*
```

repo id	repo name	status
oracle_mysql8	Oracle MySQL 8	enabled

下面用 dns install 命令安装 rpm 文件的方法生成 repo 文件。

打开页面 https://dev.mysql.com/downloads/repo/yum/,此页面可以下载各 RHEL 版本 MySQL 软件的 yum 源的 rpm 文件,在 RHEL 8 部分,单击"download"按钮,在新的页面,右击链接"No thanks, just start my download.",复制链接。

执行 dnf -y install 命令,在后面贴入复制的链接。

```
# dnf -y install https://dev.mysql.com/get/mysql80-community-release-el8-1.noarch.rpm
```

执行以上命令会在/etc/yum.repos.d 目录下创建两个 repo 文件。

```
[root /etc/yum.repos.d 2021-02-21 21:42:47]
# ls
mysql-community.repo    mysql-community-source.repo    oracle-linux-ol8.repo    uek-ol8.repo
```

5.5.3 管理软件包

(1) 列出软件包

列出所有软件包,包括开启的 yum 源提供的软件包,以及已安装的软件包。

```
# dnf list
```

显示结果分为 Installed Packages(已安装)和 Available Packages(可用)两部分。每行又分为 3 列,第 1 列为软件包名称,第 2 列为版本号或包含此软件包的模块名称,第 3 列为 yum 源(repo ID)。

可以使用通配符列出软件包,下面命令列出名称以 mysql 起始的软件包。

```
# dnf list  mysql*
Last metadata expiration check: 0:11:54 ago on Fri 26 Feb 2021 03:52:47 PM CST.
Installed Packages
mysql-community-client.x86_64          8.0.23-1.el8                              @oracle_mysql8
…
Available Packages
mysql.x86_64                           8.0.13-1.module+el8+5199+1dce7bb2    ol8_appstream
…
```

查看名称以 mysql 起始的未安装可用软件包。

```
# dnf list available mysql*
```

查看已安装的、名称以 mysql 起始的软件包。

```
# dnf list installed mysql*
```

(2) 安装、升级、删除软件包

dnf 的 install、upgrade 和 remove 选项分别用于安装、升级和删除软件包。如下面 3 个命令对 wget 软件包执行安装、升级和删除。附加-y 选项,不再要求用户确认。

```
# dnf -y install wget
# dnf -y upgrade wget
# dnf -y remove wget
```

(3) 查看安装软件包后产生的文件

查看安装某个软件包后产生的文件,使用 dnf repoquery -l 命令,需给出软件包完整名称,如查看安装 mysql-community-server 后产生的文件。

```
# dnf repoquery -l mysql-community-server
```

5.5.4 dnf 模块管理

模块是软件包的集合,一般由 Oracle Linux 系统的 Application Stream(AppStream) yum 源提供。若一个软件包同时存在系统模块和第三方 yum 源(一般是软件官方 yum 源),则优先使用模块安装软件包。若需要由第三方 yum 源安装,则要先关闭相应模块,再执行 dnf install 安装命令。

管理模块时,使用 dnf module 命令。

(1) 列出模块信息

列出所有模块信息,可以执行下面的命令。

```
# dnf module list
```

下面命令列出 postgresql 模块的信息。

```
# dnf module list postgresql
Last metadata expiration check: 0:08:47 ago on Tue 23 Feb 2021 07:49:26 AM CST.
Oracle Linux 8 Application Stream (x86_64)
Name          Stream        Profiles              Summary
postgresql    9.6           client, server [d]    PostgreSQL server and client module
postgresql    10 [d]        client, server [d]    PostgreSQL server and client module
postgresql    12            client, server [d]    PostgreSQL server and client module

Hint: [d]efault, [e]nabled, [x]disabled, [i]nstalled
```

结果中 4 个列的含义如下。

Name:模块名称。

Stream:模块版本,附带[d]为默认。如上面的示例,postgresql 模块存在 3 个版本:9.6、10 及 12。若安装模块时未指定版本,则选择默认项。

Profiles:模块中的软件包,附带[d]为默认,[i]为已安装。如上面的示例,每个 postgresql 版本的模块都存在 client 和 server 两个软件包。若安装模块时未指定此项,则选择默认项。

Summary:对模块的简要说明。

(2) 开启或关闭模块

开启或关闭模块使用 dnf module enable/disable 命令。

关闭模块后,由此模块安装的软件包会删除;安装相关软件包时,若使用 dnf install,则选用第三方提供的 yum 源,使用 dnf module install 命令则不受模块关闭的影响。若无第三方 yum 源,模块关闭后,不能再使用 dnf install 命令安装相关软件包。

关闭模块时,会关闭所有版本的模块,虽可指定版本号,但无效。

可以指定版本号开启模块。安装相关软件时,若未指定版本号,则默认安装开启的版本,而不是标记为[d]的默认版本。

若开启某个版本模块时,另外一个版本已经处于开启状态,则需要先执行 reset 命令对模块重置,再执行开启操作。

先重置 postgresql,再开启 postgresql:9.6,然后将 postgresql 模块关闭。

```
# dnf module reset postgresql
# dnf module enable postgresql:9.6
# dnf module disable postgresql
```

(3) 安装模块

安装 postgresql:9.6 的客户端软件包。

```
# dnf module install postgresql:9.6/client
```

(4) 删除安装的模块

删除已安装的 postgresql 软件包。

```
# dnf module remove postgresql
```

5.6 安装与删除 MySQL 服务器软件

在 Oracle Linux 8 系统上安装 MySQL 软件，最简单的方式是使用 dnf 工具。

5.6.1 检查 MySQL 软件是否已安装

安装之前，确认系统中未安装 MySQL 相关软件包。

```
# dnf list installed mysql*
mysql80-community-release.noarch        el8-1              @@commandline
```

以上结果是 MySQL 官方的 repo 文件，与 MySQL 软件本身无关。

5.6.2 使用 Oracle Linux 系统附带软件包安装

这是安装 MySQL 软件最简单的方式，只需一个命令，但安装的版本一般不是最新的。Oracle Linux 8 之前的版本不支持这种安装方式。

若采用模块安装方法，可以执行下面的命令。

```
# dnf -y module install mysql
```

若采用软件包安装方法，则执行下面的命令。

```
# dnf -y install mysql-server
```

mysql-server 也可以替换为@mysql，其中的@符号，表示 mysql 为一个模块。

以上命令执行完毕后，查看 mysql 模块的情况，可以得知 mysql server 已安装完成。

```
# dnf module list mysql
Oracle Linux 8 Application Stream (x86_64)
Name            Stream              Profiles                    Summary
mysql           8.0 [d][e]          client, server [d][i]       MySQL Module

Hint: [d]efault, [e]nabled, [x]disabled, [i]nstalled
```

也可以查看 mysql 相关软件包的情况，确认 mysql 相关软件都已安装。

```
# dnf list installed mysql*
Installed Packages
mysql.x86_64            8.0.21-1.module+el8.2.0+7793+cfe2b687    @ol8_appstream
mysql-common.x86_64     8.0.21-1.module+el8.2.0+7793+cfe2b687    @ol8_appstream
mysql-errmsg.x86_64     8.0.21-1.module+el8.2.0+7793+cfe2b687    @ol8_appstream
mysql-server.x86_64     8.0.21-1.module+el8.2.0+7793+cfe2b687    @ol8_appstream
```

可以看到，安装的 MySQL 版本为 8.0.21，yum 源来自操作系统的 AppStream。

下面继续说明如何管理上面过程中安装的 MySQL 服务。

5.6.3 管理 MySQL 服务

RHEL 8 系统使用 systemctl 工具管理服务，有下面两组常用选项。

- status、start、stop、restart：分别用于查看、启动、停止、重启服务。
- enable、disable：分别用于开启和关闭服务的自动启动。

执行 systemctl status 命令查看 mysqld 服务的运行状态，可看到 mysqld 未启动，而且

mysqld 未设置为自动启动。

```
# systemctl status mysqld
● mysqld.service - MySQL 8.0 database server
   Loaded:            loaded(/usr/lib/systemd/system/mysqld.service;disabled;vendor
                      preset:disabled)
   Active: inactive (dead)
```

执行 systemctl start 命令,启动 mysqld。

```
# systemctl start mysqld
```

MySQL 服务器的管理账号为 root,以 AppStream 方式安装,root 用户的口令会设置为空,可用以下方式登录 MySQL 服务器。

```
# mysql
Welcome to the MySQL monitor.  Commands end with ; or \g.
Your MySQL connection id is 8
Server version: 8.0.21 Source distribution
...
mysql>
```

查看当前会话使用的用户名,可以看到默认连接用户为 root。localhost 表示只能从本地连接 MySQL 服务器。

```
mysql> select user();
+----------------+
| user()         |
+----------------+
| root@localhost |
+----------------+
1 row in set (0.00 sec)
```

关闭 mysqld 服务,可以执行下面的命令。

```
# systemctl stop mysqld
```

设置 mysqld 随操作系统自动启动,反之,则可以使用 disable 选项。

```
# systemctl enable mysqld
```

验证结果。

```
# systemctl is-enabled mysqld
enabled
```

5.6.4 删除 MySQL 软件

若删除以上安装,可以执行以下几个命令。

删除可执行文件。

```
# dnf -y module remove mysql
```

数据目录和日志目录需要手动删除。

```
# rm -rf /var/lib/mysql
# rm -rf /var/log/mysql
```

5.6.5 使用 MySQL 官方 yum 源安装

以这种方式安装 MySQL 软件,Oracle Linux 8 只需 3 个命令(执行之前,先删除之前的安装)。

① 安装 MySQL 官方 yum 源。

```
# dnf -y install https://dev.mysql.com/get/mysql80-community-release-el8-1.noarch.rpm
```

② 禁用 Oracle Linux 系统 AppStream 源中的 mysql 模块。

```
# dnf -y module disable mysql
```

③ 安装 MySQL 软件。

```
# dnf -y install mysql-community-server
```

与 5.6.3 节的方法相同,启动 mysqld 服务。

```
# systemctl start mysqld
```

以官方 yum 源安装 MySQL 软件并启动 mysqld 服务后,会自动生成 root 账号的随机密码,存放在日志文件 /var/log/mysqld.log 中。

用 grep 命令过滤出日志文件中包含字符串"password"的行。

```
# grep password /var/log/mysqld.log
2021-01-21T13:49:07.536741Z 6 [Note] [MY-010454] [Server] A temporary password is generated for root@localhost: ug+f5oY.k<mY
```

使用以上口令,启动 mysql 客户端工具,以下面的方式连接至 mysqld 服务。

```
# mysql -uroot -p'ug+f5oY.k<mY'
```

进行其他操作前,需要重设 root 账号的口令。

```
mysql> alter user root@localhost identified by 'Root@1995';
```

若删除以上安装,可以执行以下几个命令。

删除可执行文件。

```
# yum -y remove mysql-community-server
```

手动删除数据目录和 mysqld.log 日志文件。

```
# rm -rf /var/lib/mysql
# rm -rf /var/log/mysqld.log
```

5.7 升级 MySQL

升级前,需已安装 MySQL 官方 yum 源并关闭 Oracle Linux 8 AppStream 的 mysql 模块。

```
# dnf module disable mysql
```

若旧版本以 MySQL 官方 yum 源安装,则执行下面的命令升级。

```
# dnf upgrade mysql-community-server
```

若旧版本以 AppStream 模块安装,则执行下面的命令升级。

```
# dnf upgrade mysql-server
```

5.8 目录结构

使用 dnf 方式安装 MySQL 时,MySQL 安装程序会使用下面的目录。
- /usr/sbin:存放 mysqld 程序文件。
- /usr/bin:存放客户端程序文件,如 mysql、mysqlcheck、mysqldump 等。
- /var/log:存放错误日志文件 mysqld.log。
- var/lib/mysql:数据目录,存放数据库相关文件。

5.9 连接至 mysqld 服务

要操作 MySQL 数据库,先要连接至 mysqld 服务。本节我们使用 MySQL 官方提供的客户端工具 mysql 连接服务器,mysql 是一个字符界面工具。

5.9.1 连接至本地 mysqld 服务

MySQL 数据库服务器中,最高权限账号为 root,默认只能由本地登录,下面用 root 用户连接本地服务器。

在 shell 中输入 mysql,以-u 和-p 选项分别指定用户名和密码,注意-p 后不要有空格,若有空格,则后面的字符串会被看作数据库名称。

```
# mysql -uroot -p'Root@1995'
```

可以在输入-p 后,按回车键,mysql 会提示输入密码,输入的密码不会显示出来,安全性会更高。成功连接数据库后,就可以在"mysql>"提示符后输入命令了。

5.9.2 连接至远端 mysqld 服务

假定远端 MySQL 服务器的 ip 地址为 192.168.199.133,客户端 ip 地址为 192.168.199.132。两台机器应分别满足以下条件。

192.168.199.132 机器应安装了 MySQL 客户端软件。

对于服务器,应满足以下条件:

① 已启动 mysqld 服务。

② 已创建或设置可由远端登录的 MySQL 服务器用户。

③ 已关闭防火墙 firewalld,或已开放防火墙的 3306 端口,3306 为 MySQL 专用。

对于 root 用户,其默认 host 为 localhost,表示只能由本地登录,执行下面命令将 host 属性设置为"%",从而允许 root 用户可以由任意客户端连接至服务器。

```
mysql> update mysql.user set host = '%' where user = 'root';
mysql> flush privileges;
```

执行下面的命令关闭防火墙。

```
# systemctl stop firewalld
```

若开放 3306 端口,则可以执行下面的命令。

```
# firewall-cmd --add-port=3306/tcp --permanent
# firewall-cmd - reload
```

若重新关闭 3306 端口,使用--remove-port 替换以上命令中的--add-port。

以上准备工作完成后,在 192.168.199.132 执行以下命令连接至 mysqld 服务。

```
# mysql -h'192.168.199.133' -uroot -p'Root@1995'
```

5.10 mysql 工具执行常见操作

本节使用 mysql 工具执行几个常见操作。为节省篇幅,省去部分命令的显示结果和提示信息。执行命令时,需以分号作为标记。

(1) 查看所有数据库

```
mysql> show databases;
```

(2) 创建数据库

```
mysql> create database db;
```

(3) 切换数据库(此命令不需分号)

```
mysql> use db
```

(4) 查看当前连接的数据库名称

```
mysql> select database();
```

(5) 创建表并添加、查询数据

```
mysql> create table t(a int, b int);
mysql> insert into t values(1, 10), (2, 20), (3, 30);
mysql> select * from t;
```

(6) 查看当前数据库中的所有表

```
mysql> show tables;
```

(7) 执行 SQL 命令

在 mysql 中,使用分号或\G 执行 SQL 语句,使用\G 时,行转换为列,竖向显示。

(8) 执行 SQL 脚本文件

在 mysql 中使用 source 或\.执行 SQL 脚本,两者作用相同。

```
mysql> source scott.sql
mysql> \. scott.sql
```

(9) 查询表的结构

执行 describe 命令(可简写为 desc)查询表的结构。

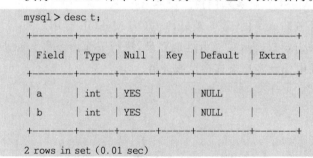

```
2 rows in set (0.01 sec)
```

(10) 修改执行过的 SQL 命令以重新执行

很多时候,要对执行过的命令进行简单修改后重新执行,可以使用上、下方向键把执行过的命令回显出来,然后使用左、右方向键移动光标到合适位置来修改特定内容。

除上述方法外,在 mysql 中可以使用 edit 命令在 vi 中打开并修改执行过的上一条 SQL 命令。修改后保存退出,使用\p 命令可将其重显出来,输入分号可重新执行。

(11) 设置默认的用户名和密码

在操作系统用户的主目录下,创建.my.cnf 文件,将连接 mysqld 服务的用户名及其密码按下面的形式写入其中,再次使用 mysql 或其他客户端工具登录 mysqld 服务时,不需输入用户名和密码。

```
# cat /root/.my.cnf
[client]
user = root
password = Root@1995
```

(12) 自定义 mysql 提示符

使用 prompt 可以自定义 mysql 提示符,常用的几个参数包括:\u(表示用户名)、\d(表示数据库名称)、\R(表示 24 小时制的小时)、\m(表示分钟),\s(表示秒)。

修改提示符,使其包含当前用户名和数据库名称。

```
mysql> prompt SQL[\u@\d]>
PROMPT set to 'SQL[\u@\d]>'
SQL[root@(none)]> use db
Database changed
SQL[root@db]>
```

(13) 获得 mysql 帮助信息

启动 mysql 时,输入--help,可以得到其帮助信息。

```
# mysql --help
```

5.11 创建测试数据

本书测试数据为一个人力资源系统使用的 3 张表,emp、dept 及 salgrade,分别表示员工表、部门表和工资级别表,比 MySQL 提供的测试数据简单。下面是创建测试数据的 SQL 脚本文件的内容,其名称为 scott.sql,改编自 Oracle 数据库的同名测试脚本文件。

```
# cat scott.sql
-- 创建 emp, dept, salgrade 测试表
drop table if exists sales, emp, dept, salgrade
;
create table dept
(
    deptno      int,
    dname       varchar(14),
    loc         varchar(13),
    primary key(deptno)
)
;
```

```sql
create table emp
(
    empno      int,
    ename      varchar(10),
    job        varchar(9),
    mgr        int,
    hiredate   date,
    sal        numeric(7,2),
    comm       numeric(7,2),
    deptno     int,
    primary key(empno),
    constraint fk_deptno foreign key(deptno) references dept(deptno)
)
;
create table salgrade
(
    grade    int not null,
    losal    int not null,
    hisal    int not null
)
;
insert into dept values
(10,'ACCOUNTING','NEW YORK'),
(20,'RESEARCH','DALLAS'),
(30,'SALES','CHICAGO'),
(40,'OPERATIONS','BOSTON')
;
insert into emp values
(7369,'SMITH','CLERK',7902,'1980-12-17',800,NULL,20),
(7499,'ALLEN','SALESMAN',7698,'1981-02-02',1600,300,30),
(7521,'WARD','SALESMAN',7698,'1981-02-22',1250,500,30),
(7566,'JONES','MANAGER',7839,'1981-04-02',2975,NULL,20),
(7654,'MARTIN','SALESMAN',7698,'1981-09-28',1250,1400,30),
(7698,'BLAKE','MANAGER',7839,'1981-05-01',2850,NULL,30),
(7782,'CLARK','MANAGER',7839,'1981-06-09',2450,NULL,10),
(7839,'KING','PRESIDENT',NULL,'1981-11-17',5000,NULL,10),
(7844,'TURNER','SALESMAN',7698,'1981-09-08',1500,0,30),
(7900,'JAMES','CLERK',7698,'1981-12-03',950,NULL,30),
(7902,'FORD','ANALYST',7566,'1981-12-03',3000,NULL,20),
(7934,'MILLER','CLERK',7782,'1982-01-23',1300,NULL,10)
;
insert into salgrade values
(1, 700, 1200),
```

```
(2, 1201, 1400),
(3, 1401, 2000),
(4, 2001, 3000),
(5, 3001, 9999)
;
```

第 6 章　SQL 查询语句

　　SQL 语言是操作关系型数据库的标准语言。和其他编程语言相比，SQL 语言与日常英语更接近，比较容易掌握。从事数据库相关工作需要具备快速编写 SQL 语句的能力。本章主要讲述 SQL 语言的核心部分——查询。

　　本章主要内容包括：
- SQL 语言的历史；
- 常用数据类型；
- 简单的 SQL 查询语句；
- 常用函数；
- 汇总函数与子查询；
- 集合操作；
- 表连接查询；
- 构造复杂的查询语句；
- SQL 查询的等效转换。

6.1　SQL 概述

　　从 1973 年 IBM 开始研发 SQL 语言至今，已经过去近 50 年，SQL 的功能不断增强，SQL 标准的篇幅，从 SQL-86 的 100 多页扩充至现在的几千页。

6.1.1　SQL 语言的历史

　　1970 年，Codd 提出了关系模型。他构造了操作关系的两种面向集合语言——关系代数和关系演算。这两种语言以严谨的数学符号表述，不能在键盘上输入，也很难看懂，距离实际应用还有很长一段路。

　　1973 年，Don Chamberlin 和 Ray Boyce 在 System R 项目中负责开发数据库操作工具。其任务是把 Codd 的数学语言变为贴近日常英语、容易理解的形式，同时其语法结构又要足够严谨，使计算机容易解析。

　　1974 年，操作数据的 DML 与定义数据对象的 DDL 基本成形了，人们把这种语言命名为 SEQUEL，即 Structured English Query Language，这里的 Structured 指其语法结构严格，各子句有固定顺序。SEQUEL 本意为查询语言，但查询功能只是其主要部分，除此之外，它还具备增删、修改数据，创建或修改数据库对象等功能。

　　1977 年，项目组发现 SEQUEL 已被英国飞机公司 Hawker Siddeley 用作商标，只得将其重命名为 SQL，意为 Structured Query Language。

　　20 世纪 70 年代末，包括 Larry Ellison 在内的 Oracle 公司的 3 个创始人，参考 IBM 公司公开的 SQL 资料,开发自己的关系型数据库产品,并于 1979 年推出了世界上第一个关系型数据库产品——Oracle V2,其操作语言也使用 SQL,并遵循 IBM 的语法规则。

6.1.2 SQL 的发音

按首版 SQL 标准,SQL 的发音为 S-Q-L,即[es kju: el],但包括 Don Chamberlin 本人在内,人们一般更习惯地将其读为最初的"sequel",即[' si:kw(ə)l]。

6.1.3 SQL 查询的特点

SQL 查询的最显著特点是面向集合。使用 SQL 语言执行查询时,只需说明数据满足的条件,不必说明数据的存储位置,也不用说明如何遍历数据。

6.1.4 SQL 标准

随着关系型数据库的普及,SQL 逐渐成为其通用语言。为了系统应用可以在不同数据库产品之间顺利移植,ANSI(美国标准化局)在 1986 年组织制定了第一个 SQL 标准,这个标准于 1987 年被 ISO(国际标准化组织)接受,成为第一个 SQL 国际标准,一般称为 SQL-86。

数据库的功能在不断扩充,SQL 标准也在不断更新,ISO 陆续推出了 SQL-86、SQL-89、SQL-92、SQL:1999、SQL:2003、SQL:2006、SQL:2008、SQL:2011、SQL:2016 等版本。

当不同厂商实现 SQL 语言时,会尽量使其符合国际标准。用户在熟悉一个厂商的 SQL 语言后,所获得的经验可以迁移至其他厂商的 SQL 语言,不必再从头学起。

6.2 SQL 语言的主要类型

SQL 语言可以完成多种功能,其最常用的功能为查询及修改数据,创建数据库对象,权限管理。在 SQL-92 中,这三类语句划分为 DML、DDL 及 DCL(数据控制语言,Data Control Language),主要 SQL 语句和功能如表 6-1 所示。

表 6-1 SQL 语言分类

分类	SQL 语句	功能
DML	select	查询表的数据
	update	修改表中的列值
	delete	删除表中的行
	insert	对表添加新行
DDL	create	创建数据库对象,如表、视图、索引等
	alter	修改数据库对象定义,如对表添加或删除列
	drop	删除数据库对象
	truncate	以释放表的空间的方式清空表的数据
DCL	grant	对用户赋予权限
	revoke	撤销用户的权限

在以上 SQL 语句中,最常用、最复杂的是查询语句,本章将由简到繁逐步讲解 SQL 查询

的各种用法,后续章节将讲解 DML 语句中的数据修改部分和 DDL 语句。

6.3 常用数据类型

创建表时,需要指定列的数据类型,查询或修改表中的数据时,要根据其数据类型采用合适的操作方式。本节介绍 MySQL 数据库的 3 种常用数据类型:数值、字符串和日期时间。

6.3.1 数值类型

数值类型主要包括整型、定点小数及浮点数。

对于整型,MySQL 主要支持 int 及 bigint,分别使用 4 字节和 8 字节存储。另外,MySQL 也支持 tinyint、smallint 及 mediumint,以满足更精细的需要。

MySQL 的定点小数类型用 decimal(p, s) 或 numeric(p, s) 表示,二者同义,p 表示有效数字位数,s 表示小数位数。p 和 s 符合以下要求:$1 \leqslant p \leqslant 65, 0 \leqslant s \leqslant 30, s \leqslant p$。若省略参数,则表示整数,$p$ 默认为 10。若省略 s,则 s 默认为 0。

对于浮点型数值,MySQL 支持 float 和 double 类型。浮点数的表示和运算均存在误差,除非必要,数据库一般不使用浮点数。

6.3.2 字符串类型

字符串类型包括 char(n) ($n \leqslant 255$) 和 varchar(n) ($n \leqslant 65\,535$),n 为允许的最大长度,单位为字符。字符串实际长度小于 n 时,char(n) 会用空格填满,而 varchar(n) 所占空间以实际字符数为准。使用字符串常量时,需要用单引号或双引号括住,如'abc'。

MySQL 8 默认使用 utf8mb4 字符集,不再存在字符包含范围方面的问题。

字符串排序时,需要指定规则,如汉字以拼音排序,英文字母是否区分大小写。字符串的排序规则可以在服务器、数据库、表或列等多个层次规定。

6.3.3 日期时间类型

日期时间类型主要包括以下几种。

year:表示年份。范围为 1901~2155,以 1 字节存储。

date:表示日期。范围为 1000-01-01~9999-12-31,以 3 字节存储。

time(p):表示时间。范围为 -838:59:59~838:59:59,以 3 字节存储。除了表示时刻,也用于记录时间段长度,以及两个时刻的差值。

datetime(p):表示日期时间。范围为 1000-01-01 00:00:00~9999-12-31 23:59:59,以 8 字节存储。

timestamp(p):表示包含时区信息的日期时间。范围为 1970-01-01 00:00:01 UTC~2038-01-19 03:14:07 UTC,以 4 字节存储。客户端的当前时区为服务器时区,存储时,会转换为 UTC(Universal Time Coordinated)时间。

p 参数表示秒的小数部分精确度,最大为 6,默认为 0。

一般以规定格式的字符串表示日期时间常量,如'2021-02-12 16:30:00'。

6.4 简单的 SQL 查询语句

SQL 语言使用 select 语句执行查询操作,其简单形式一般由 3 个子句构成。
- select 子句指定列名。
- from 子句指定表名。
- where 子句指定查询条件,借鉴了英语定语从句的用法。

6.4.1 最简单的查询——只指定表

要得到一个表的所有数据,只需要在 select 子句中使用"*",在 from 子句中指定表名。如查询 dept 表的所有数据,可以执行以下语句。

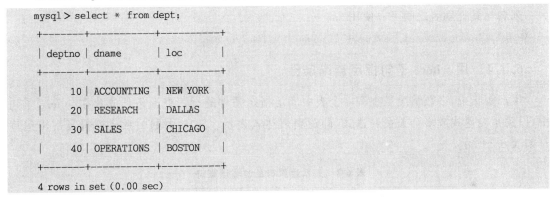

6.4.2 指定列

若只查询若干列,可以在 select 子句中指定列名。如查询 dept 表中,每个部门的名称与地址,可以执行以下命令。

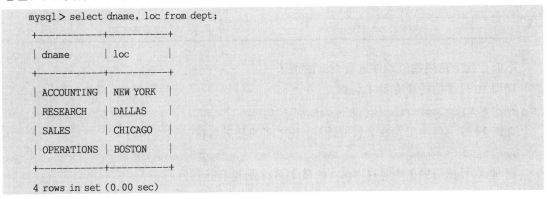

6.4.3 指定列别名

若列名的含义不清晰,可以使用列别名提高查询结果的可读性。如查询 dept 表中,每个部门的名称与地址,使用列别名 DEPT NAME 和 LOCATION 代替 dname 和 loc。

```
mysql> select dname "DEPT NAME", loc LOCATION from dept;
+------------+----------+
| DEPT NAME  | LOCATION |
+------------+----------+
| ACCOUNTING | NEW YORK |
| RESEARCH   | DALLAS   |
| SALES      | CHICAGO  |
| OPERATIONS | BOSTON   |
+------------+----------+
4 rows in set (0.00 sec)
```

因为 DEPT 与 NAME 之间有空格,将其作为列别名使用时,要加双引号,如果列别名恰好是 SQL 关键字,也要加上双引号。

列名与其列别名之间可以使用 as。

```
mysql> select dname as "DEPT NAME", loc as LOCATION from dept;
```

6.4.4 用 where 子句指定查询条件

生产实际中,多数情况是查询一个表中满足指定条件的行。查询条件放置在 where 子句中,以关系表达式或多个关系表达式,即逻辑表达式表示。SQL 中的关系运算符及逻辑运算符如表 6-2 所示。

表 6-2 关系运算符及逻辑运算符

运算符	功能
=,<,>,<>或!=,<=,>=,between…and	大小比较
in,exists	子查询中是否存在指定值
like	字符串模糊比较
is null,is not null	空值判断
and,or,not 或!	逻辑运算符

下面是几个使用数值条件进行查询的例子。

查询 10 号部门的所有员工名称。

```
mysql> select ename from emp where deptno = 10;
```

查询 10 号部门中工资在 2 000 到 3 000 之间的员工名称。

```
mysql> select ename, sal from emp where deptno = 10 and sal between 2000 and 3000;
```

查询 10 号部门中工资超过 2 000 的员工名称及其工资。

```
mysql> select ename, sal from emp where deptno = 10 and sal > 2000;
```

涉及字符串和日期时间的条件查询较复杂,后面单独说明。

6.4.5 使用 order by 子句给查询结果排序

查询结果是行的集合,本来不存在前后顺序,为了查看方便,加入 order by 子句可让用户按指定列对查询结果排序。

如查询员工名称和工资,要求按照工资从低到高升序显示查询结果。

```
mysql> select ename, sal from emp order by sal asc;
```
asc 是默认值，可以省略。asc 对应单词 ascend(上升)。
若按工资降序排列显示，则附加 desc 关键字。desc 对应单词 descend(下降)。
```
mysql> select ename, sal from emp order by sal desc;
```
order by 子句中可以指定多个列排序，每个列可以单独指定排序方式(asc 或 desc)。如按两个列排序，先把查询结果按照第 1 个列排序，如果第 1 个列中有重复值，则这些行再按照第 2 个列排序，若第 1 个列中没有重复值，则第 2 个排序列不起作用。若包含 3 个以上的排序列，处理方式与此相同。

把查询结果按照 deptno 和 sal 排序。
```
mysql> select deptno, ename, sal from emp order by deptno,sal;
```
除了可以在 order by 子句中使用列名指定排序列，也可以使用列别名。
```
mysql> select ename, sal salary from emp order by salary asc;
```
还可以使用排序列在 select 子句中出现的序号。
```
mysql> select deptno, ename, sal from emp order by 1, 3;
```

6.4.6　分页查询

MySQL 诞生之初就和互联网应用紧密结合，当时的一类典型应用是论坛，把大量的帖子标题分页显示是其必备功能。MySQL 在 1995 年就使用 limit m,n 子句实现了简洁的分页功能，其功能是略过前面 m 行，再顺序取出 n 行。若省略 m，则 m 默认为 0。

以下查询能够得到最高工资的第 3 至第 5 名。
```
mysql> select ename, sal from emp order by sal desc limit 2, 3;
```
为了与 PostgreSQL 兼容，MySQL 也支持下面的语法(m,n 的顺序相反)。
```
mysql> select ename, sal from emp order by sal desc limit 3 offset 2;
```

6.5　常用数值运算符及函数

MySQL 提供了丰富的函数处理数据。本节讲解常用的数值函数。

6.5.1　数值运算符

(1) +，-，*

这 3 个运算符即普通的加减乘运算，若整数(int 或 bigint)参与运算，则返回值为 bigint。

(2) /，div，%

/为普通除法，商为小数；div 为整数除法，商为整数；%为求余数，与 mod()功能相同。

6.5.2　常用数值函数

- abs()：abs()用于求绝对值。
- ceiling()：返回不小于参数的最大整数。
- floor()：返回不大于参数的最小整数。
- round(x,n)，truncate(x,n)：若省略 n，则默认为 0。若 $n>0$，则 round(x, n)经四舍五入后，保留 n 位小数。若 $n<0$，则把整数部分四舍五入至左向第 $n+1$ 位(个位数看作

第 1 位),后面 n 位均取值为 0。

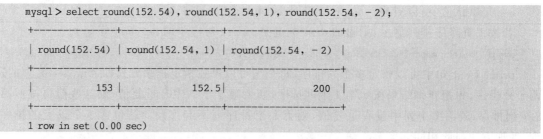

truncate(x,n)把 x 截断至 n 位小数,若 $n<0$,则把小数点左侧的 n 位取值为 0。

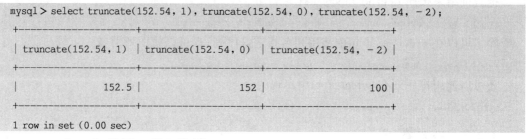

- power(m,n):返回 m 的 n 次幂。
- rand():返回一个 0 至 1 范围内的随机浮点数。

rand()函数可以应用于平时的随机抽奖活动中,如下示例表示随机抽取 emp 表中的 3 个员工作为获奖人员。

```
mysql> select empno, ename from emp order by rand() limit 3;
```

除了这几个常用函数外,还有三角函数、对数函数、指数函数等。

6.6 字符数据的处理

数据库中的主要数据类型包括数值、字符串及日期,本节讲解字符串处理。

6.6.1 字符串常量

字符串常量需要加单引号或双引号。英文字母是否区分大小写,由排序规则确定。
查询 emp 表中,SMITH 的工资值。

```
mysql> select ename,sal from emp where ename = 'SMITH';
```

字符串也可以比较大小。

```
mysql> select ename,sal from emp where ename >'SMITH';
```

6.6.2 字符串模糊查询

很多情况下,用户可能会忘记字符串常量的完整写法,这时可以使用字符串模糊查询。
执行字符串模糊查询时,使用"like"关键字,并以"%"或"_"作为通配符。
- %:表示任意个任意字符。
- _:表示 1 个任意字符。

下面通过几个示例说明字符串模糊查询的用法。
查询 emp 表中,以字母 S 开头的员工名称。

```
mysql> select ename from emp where ename like 'S%';
```
查询 emp 表中,第 2 个字母是 A 的员工名称。
```
mysql> select ename from emp where ename like '_A%';
```
查询 emp 表中,包含字母 S 的员工名称。
```
mysql> select ename from emp where ename like '%S%';
```

6.6.3 处理字符串中的特殊字符

有时候需要查询包含某些特殊字符的字符串,而这些字符在 SQL 中又有固定意义,如通配符"%"和"_",以及单引号、双引号。

查询包含"%"和"_"的字符串,需要在其前面使用转义符,MySQL 限定使用反斜杠"\"作为转义符。

查询包含"_"的部门名称,以及包含"%"的部门地址。
```
mysql> select * from dept where dname like '%\_%';
mysql> select * from dept where loc like '%\%%';
```
在 SQL 中,单引号是作为字符串常量定界符使用的,如果字符串中的单引号作为真正字符,而不是定界符,可以使用两个单引号,此时第 1 个单引号起转义作用,表示第 2 个单引号作为真正字符使用。
```
mysql> select 'McDONALD''S';
```
MySQL 也支持使用双引号作为字符串定界符,这时,单引号被看作普通字符。下面的查询使用了这种处理方法。
```
mysql> select "McDONALD'S";
```
类似地,单引号内的双引号被看作普通字符。
```
mysql> select 'McDONALD"S';
```
另外,MySQL 也支持使用反斜杠"\"对定界符转义。
```
mysql> select 'McDONALD\'S', "McDONALD\"S";
```
在转义符前面使用转义符,可以把"\"自身转义为普通字符。
```
mysql> select 'McDONALD\\S';
```

6.6.4 常用字符串函数

MySQL 提供了丰富的函数用于处理字符串。

(1) char_length(),length()

返回字符串的长度。char_length()返回字符数,length()返回字节数。

在 utf8mb4 字符集中,ASCII 码字符的字节数与字符数是相同的,对于其他文字,如汉字,字符数一般小于字节数。

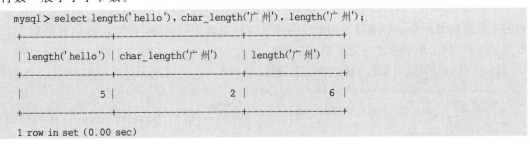

(2) concat()、concat_ws()

两个函数的作用是合并字符串。concat_ws()可以在字符串之间指定分隔符，ws 表示 with separator。concat_ws()的第 1 个参数为分隔符，其他是要合并的字符串。

```
mysql> select concat('Hello','My','SQL'), concat_ws(',','Hello','My','SQL');
+----------------------------+-----------------------------------+
| concat('Hello','My','SQL') | concat_ws(',','Hello','My','SQL') |
+----------------------------+-----------------------------------+
| HelloMySQL                 | Hello,My,SQL                      |
+----------------------------+-----------------------------------+
1 row in set (0.01 sec)
```

如果把 sql_mode 设置为 pipes_as_concat，"||"可以用作字符串合并运算符，"||"也是 SQL 标准的运算符。

```
mysql> set sql_mode = 'pipes_as_concat';
Query OK, 0 rows affected (0.00 sec)

mysql> select 'Hello,'||'My'||'SQL';
+-----------------------+
| 'Hello,'||'My'||'SQL' |
+-----------------------+
| Hello,MySQL           |
+-----------------------+
1 row in set (0.00 sec)
```

set sql_mode ＝ 'pipes_as_concat'；只对当前连接有效，更多内容请参考系统参数设置相关章节。

(3) lower()、upper()

这两个函数把字符串转换为小写/大写。

(4) lpad()、rpad()

在指定字符串左/右侧附加指定字符，使字符串达到指定长度。

```
mysql> select lpad('MySQL', 10,'#'), rpad('MySQL', 10,'#');
+-----------------------+-----------------------+
| lpad('MySQL', 10,'#') | rpad('MySQL', 10,'#') |
+-----------------------+-----------------------+
| #####MySQL            | MySQL#####            |
+-----------------------+-----------------------+
1 row in set (0.00 sec)
```

不需借助其他工具，使用这两个函数可实现简单的数字可视化效果，下面的示例把 emp 表的工资值高低用 lpad()函数返回的字符串长度形象地表示出来，下面 lpad 函数的第 1 个参数为空字符，两个单引号之间无空格。

```
mysql> select ename, sal, lpad('', round(sal div 100),'*') from emp;
+-------+------+----------------------------------+
| ename | sal  | lpad('', round(sal div 100),'*') |
```

```
+---------+----------+---------------------------------------------+
| SMITH   |   800.00 | ********                                    |
| ALLEN   |  1600.00 | ****************                            |
| WARD    |  1250.00 | ************                                |
| JONES   |  2975.00 | *****************************               |
| MARTIN  |  1250.00 | ************                                |
| BLAKE   |  2850.00 | ****************************                |
| CLARK   |  2450.00 | ************************                    |
| KING    |  5000.00 | **************************************      |
| TURNER  |  1500.00 | ***************                             |
| JAMES   |   950.00 | *********                                   |
| FORD    |  3000.00 | ******************************              |
| MILLER  |  1300.00 | *************                               |
+---------+----------+---------------------------------------------+
12 rows in set (0.00 sec)
```

(5) trim()、ltrim()、rtrim()

ltrim()、rtrim()表示分别去除字符串左右侧的空格。

```
mysql> select ltrim('  MySQL  '), rtrim('  MySQL  ');
```

trim()的功能比以上两个函数丰富,可以使用 leading、trailing、both 参数,去除左右单侧,或两侧的指定字符,默认去除空格。

```
mysql> select trim('  MySQL  ') "default",
    ->        trim(leading 'x' from 'xxMySQLxx') "trim_leading_x",
    ->        trim(trailing 'x' from 'xxMySQLxx') "trim_trailing_x",
    ->        trim(both 'x' from 'xxMySQLxx') "trim_both_x";
+---------+----------------+-----------------+-------------+
| default | trim_leading_x | trim_trailing_x | trim_both_x |
+---------+----------------+-----------------+-------------+
| MySQL   | MySQLxx        | xxMySQL         | MySQL       |
+---------+----------------+-----------------+-------------+
1 row in set (0.01 sec)
```

(6) replace()

用子串替换字符串中的指定部分。下面的示例把字符串中的 World 替换为 MySQL。

```
mysql> select replace('Hello, World', 'World', 'MySQL');
+-------------------------------------------+
| replace('Hello, World', 'World', 'MySQL') |
+-------------------------------------------+
| Hello, MySQL                              |
+-------------------------------------------+
1 row in set (0.00 sec)
```

(7) repeat()、space()

repeat(e,n)把字符串 e 重复 n 次,space(n)返回由连续 n 个空格构成的字符串。

```
mysql> select repeat('#', 8) "repeat", concat('hello,', space(5), 'world') "space";
+----------+--------------------+
| repeat   | space              |
+----------+--------------------+
| ######## | hello,     world   |
+----------+--------------------+
1 row in set (0.00 sec)
```

(8) locate(),instr(),position()

3个函数都返回特定子串在一个字符串中的位置,都包含子串和字符串2个参数,语法稍有不同。2个参数中,instr()函数要求子串在后,locate()和position()函数要求子串在前,position()在两参数之间使用in。

```
mysql> select instr('Oracle','RAC') as "instr",
    ->        locate('RAC','Oracle') as "locate",
    ->        position('RAC' in 'Oracle') as "position";
+-------+--------+----------+
| instr | locate | position |
+-------+--------+----------+
|     2 |      2 |        2 |
+-------+--------+----------+
1 row in set (0.00 sec)
```

(9) substring_index(*str*, *d*, *n*)

3个参数中,*str*为字符串,*d*为分隔符,*n*为分隔符序号。此函数返回*str*第*n*个分隔符之前的部分,若*n*为负数,分隔符计数由右侧开始,截取的子串也是分隔符右侧的部分。

```
mysql> select substring_index('www.mysql.com','.', 2),
    ->        substring_index('www.mysql.com','.', -2);
+-----------------------------------------+------------------------------------------+
| substring_index('www.mysql.com','.', 2) | substring_index('www.mysql.com','.', -2) |
+-----------------------------------------+------------------------------------------+
| www.mysql                               | mysql.com                                |
+-----------------------------------------+------------------------------------------+
1 row in set (0.00 sec)
```

(10) insert(*str*, *pos*, *len*, *substr*)

此函数的功能是把一个子串插入字符串,返回新字符串。*str*是原字符串;*substr*是要插入的子串;*pos*是插入的起始位置;*len*为替换掉的字符个数,*len*为0则不替换。

```
mysql> select insert('12345', 2, 0, 'AAA'), insert('12345', 2, 2, 'AAA');
+------------------------------+------------------------------+
| insert('12345', 2, 0, 'AAA') | insert('12345', 2, 2, 'AAA') |
+------------------------------+------------------------------+
| 1AAA2345                     | 1AAA45                       |
+------------------------------+------------------------------+
1 row in set (0.00 sec)
```

(11) left(),right()

分别返回由左侧或右侧开始，指定个数的字符串。

```
mysql> select left('abcdef', 3), right('abcdef', 3);
+-------------------+--------------------+
| left('abcdef', 3) | right('abcdef', 3) |
+-------------------+--------------------+
| abc               | def                |
+-------------------+--------------------+
1 row in set (0.00 sec)
```

(12) substr(),mid()

两个函数用法相同，均为从一个字符串中返回子串。有3个参数，第1个是要处理的原字符串；第2个是截取子串的起始位置，若为正数，则由左侧开始计数，若为负数，则由右侧开始计数；第3个是截取的子串长度，若省略第3个参数，则截取至末尾。

```
mysql> select substr('abcdef', 3), substr('abcdef', 3, 2), substr('abcdef', -3, 2);
+---------------------+------------------------+-------------------------+
| substr('abcdef', 3) | substr('abcdef', 3, 2) | substr('abcdef', -3, 2) |
+---------------------+------------------------+-------------------------+
| cdef                | cd                     | de                      |
+---------------------+------------------------+-------------------------+
1 row in set (0.01 sec)
```

6.6.5 利用正则表达式搜索字符串

正则表达式是一个描述文本模式的字符序列。经过多年发展，其语法规则已成为POSIX标准的一部分，是文本搜索的重要工具。MySQL处理正则表达式的函数主要有regexp_like()、regexp_substr()、regexp_replace()、regexp_instr()。本节以regext_like()函数为例，简单介绍一下MySQL的正则表达式用法。

正则表达式中常用的元字符有以下几个。

*：表示*符号前面的字符可以包含0个或任意个，注意与文件名称中的*通配符区别。

.：表示1个任意字符。若.与*一起使用，".*"则表示任意个任意字符。

$：表示前面的字符在一行的结尾。

[]：用于指定字符范围，如[abc]表示abc 3个字符中的任意一个，[0-9]表示任意一个数字，[a-z]表示任意一个小写字母，[A-Z]表示任意一个大写字母。

^：用到方括号中，表示不包含某个范围的字符，用到一个字符串的前面，表示这个字符串出现在一行的开头。

创建测试表park。

```
mysql> create table park(
    ->     park_name varchar(40),
    ->     park_phone varchar(15),
    ->     country varchar(2),
    ->     description text
    -> );
```

添加3行测试数据。

```
mysql> set sql_mode = 'pipes_as_concat';
mysql> insert into park values
    -> (
    ->     'Mackinac Island State Park',
    ->     '(231) 436-4100',
    ->     'US',
    ->     'Michigan"s first state park encompasses approximately 1800 acres'
    ->     ||'of Mackinac Island. The centerpiece is Fort Mackinac,'
    ->     ||'built in 1780 by the British to protect the Great Lakes Fur Trade.'
    ->     ||'For information by phone, dial 800-44-PARKS or 517-373-1214.'
    -> ),
    -> (
    ->     'Laughing Whitefish Falls Scenic Site',
    ->     '(906) 863-9747',
    ->     'US',
    ->     'This scenic site is centered around an impressive waterfall.'
    ->     ||'A rustic, picnic area with waterpump is available.'
    -> ),
    -> (
    ->     'Pukaskwa National Park',
    ->     '(807) 229-0801',
    ->     'CA',
    ->     'Pukaskwa National Park is on the north shore of Lake Superior,'
    ->     || char(10) ||'and is "twinned" with the Pictured Rocks National'
    ->     || char(10) ||'Lakeshore almost directly south in Michigan.'
    ->     || char(10) ||'For information on Pukaskwa, phone 807-229-0801.'
    -> );
```

查找 description 列中包含电话号码的 park_name。

```
mysql> select park_name from park
    -> where regexp_like(description,'[0-9]{3}-[0-9]{3}-[0-9]{4}');
```

[0-9]表示 0 到 9 之间的数字，{3}表示前面的字符是 3 个，[0-9]{3}则表示 3 个数字。
[0-9]也可以表示为[:digit:]，如查询 description 列中包含数字的行。

```
mysql> select park_name from park where regexp_like(description,'.*[:digit:].*');
```

这里的".*"表示任意个任意字符。类似[:digit:]的用法还有下面几个：[:alpha:]表示字母；[:blank:]表示空白，如空格、制表符；[:lower:]/[:upper:]表示小写或大写字母。

查询包含字符"-"的行。

```
mysql> select park_name from park where regexp_like(description,'.*-.*');
```

与 like 的用法一样，如果查询条件包含了元字符或者其他特殊字符（如括号等），则可以使用"\"转义。查询 park_phone 列的第 1 个数字不是 9 的行。

```
mysql> select park_name,park_phone from park where regexp_like(park_phone,'\\([^9].*');
```

"\("表示对括号字符"("转义，要使用两个"\"，[^9]表示此位置不是字符 9。

查找 description 列中以 available.结尾的行。

```
mysql> select park_name from park where regexp_like(description,'available.$');
```

查找 description 列以字母 M 开头的行。

```
mysql> select park_name from park where regexp_like(description,'^M');
```

6.7 处理日期型数据

处理日期型数据时,一般有下面几种情况:日期常量的表示,日期值的显示,抽取日期的指定部分。这里的日期型数据也包括时间。

6.7.1 获得当前日期时间

MySQL 一般使用 now() 函数得到当前日期时间信息,可以指定秒的小数部分的精度,其返回值的类型为 timestamp,使用 utc_timestamp() 函数可以返回当前的 UTC 时间。

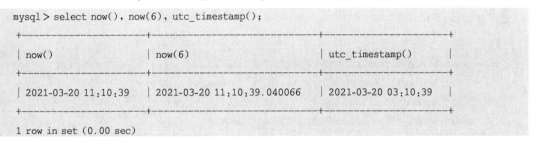

6.7.2 日期型常量

日期型常量可以用指定格式的字符串常量表示。其各部分需按顺序分别表示年份、月份和日,小时、分、秒,年份 4 位,其他部分 2 位,分隔符可自行选定,一般日期各部分以 "-" 或 "/" 分隔,时间部分以 ":" 分隔,如 'yyyy-mm-dd hh24:mi:ss'。MySQL 会把符合这种格式的字符串常量转换为日期型常量。

执行下面的命令,MySQL 自动把 '2020/08/12 08:30:00' 转换为日期型常量。

```
mysql> insert into emp(empno, ename, hiredate)
    -> values(9999,'ADAMS','2020/08/12 08:30:00');
```

也可以用 str_to_date(e, fmt) 函数把特定格式的字符串转换为日期型常量。e 为字符串,fmt 为格式码。格式码用于指定字符串 e 中的各个组成部分。

%Y 表示 4 位年份,%m 和 %d 分别表示 2 位月份和日,%H 表示 2 位 24 小时制的小时,%i 和 %s 分别表示 2 位的分和秒。如 str_to_date('08/12/2020','%m/%d/%Y') 转换字符串为 2020 年 8 月 12 日。

下面示例用 str_to_date() 函数转换字符串为日期型常量。

```
mysql> insert into emp(empno, ename, hiredate)
    -> values(8888,'SCOTT', str_to_date('08/12/2020 08:30:00','%m/%d/%Y %H:%i:%s'));
```

6.7.3 指定格式显示日期型列值

date_format() 函数与 str_to_date() 函数的功能恰好相反,它是将日期型数据转换为符合指定格式的字符串,二者使用格式码的方式相同。

```
mysql> select ename, date_format(hiredate,'%m-%d-%Y %H:%i:%s') from emp
    -> where empno = 7369;
+-------+--------------------------------------------+
| ename | date_format(hiredate,'%m-%d-%Y %H:%i:%s')  |
+-------+--------------------------------------------+
| SMITH | 12-17-1980 00:00:00                        |
+-------+--------------------------------------------+
1 row in set (0.00 sec)
```

6.7.4 抽取日期的指定部分

MySQL 提供了 extract(*unit* from *e*) 函数,从日期常量中抽取指定部分。*unit* 参数用于指定抽取的部分,*e* 为日期常量。*unit* 可以取自以下关键字:MICROSECOND、SECOND、MINUTE、HOUR、DAY、WEEK、MONTH、QUARTER、YEAR。

具体用法如下。

```
mysql> select
    -> extract(year from '2020-09-01 11:30:00') year,
    -> extract(month from '2020-09-01 11:30:00') month,
    -> extract(day from '2020-09-01 11:30:00') day,
    -> extract(hour from '2020-09-01 11:30:00') hour,
    -> extract(quarter from '2020-09-01 11:30:00') quarter;
+------+-------+-----+------+---------+
| year | month | day | hour | quarter |
+------+-------+-----+------+---------+
| 2020 |     9 |   1 |   11 |       3 |
+------+-------+-----+------+---------+
1 row in set (0.00 sec)
```

除了 extract() 函数外,MySQL 也提供了 year()、month()、day()、hour()、minute()、second()、quarter() 等函数直接抽取指定部分。还提供了 dayofyear()、weekofyear() 返回日期和周在一年里的序号,dayname()、monthname() 返回周几和月份的英文名称。

```
mysql> select
    -> year('2020-08-12 11:30:00') year,
    -> dayname('2020-08-12 11:30:00') dayname,
    -> weekofyear('2020-08-12 11:30:00') weekofyear;
+------+-----------+------------+
| year | dayname   | weekofyear |
+------+-----------+------------+
| 2020 | Wednesday |         33 |
+------+-----------+------------+
1 row in set (0.00 sec)
```

6.7.5 获取时间差

MySQL 可以使用以下函数获取两个日期常量的差值。

datediff(*e*1, *e*2):得出 *e*1−*e*2,单位是天数。*e*1 和 *e*2 为日期常量,时间部分会被忽略。

timestampdiff(*unit*, *e*1, *e*2):得出 *e*1−*e*2,单位由 *unit* 参数指定,*e*1 和 *e*2 为日期常量,若不包含时间部分,则时间部分指定为 00:00:00。unit 为以下关键字之一:MICROSECOND、SECOND、MINUTE、HOUR、DAY、WEEK、MONTH、QUARTER、YEAR。

下面的示例为计算两个日期之间相差的天数。

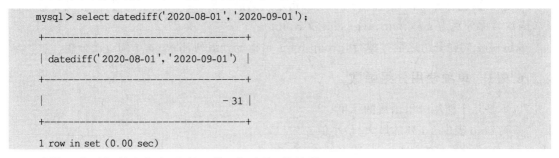

计算两个时间值之间相差的天数、小时数、分钟数。

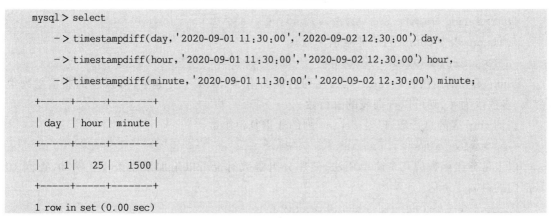

6.8 空值的处理

在对表添加记录时,若某个列因为某种原因未指定值,则称此列值为 null。若一个表达式包含 null 值,则其结果也为 null,即不确定,不确定的布尔值当作 FALSE。

查询某个列上为空或不为空的值,不能使用"="和"<>",而要使用 is null 或 is not null。

如查询 emp 表中,comm 列为空的员工名称。

```
mysql> select ename from emp where comm is null;
```

反之,查询 emp 表中,comm 列不为空的员工名称,则要使用 is not null。

```
mysql> select ename from emp where comm is not null;
```

查询 emp 表中,每个员工每个月的收入之和,即 sal+comm。若直接查询 sal+comm,当 comm 为 null 时,此和也为 null。

ifnull(*x*, *n*)函数专门处理空值,若 *x* 不为空,则返回 *x*,若 *x* 为 null,则返回 *n*。

若 comm 值为 null 时,将其看作 0,则可以执行如下查询。

```
mysql> select ename, sal, comm, sal + ifnull(comm, 0) from emp where deptno = 30;
```

6.9 分组汇总

分组汇总即把某个列上有相同值的记录各自划分为一组,然后再对各个组的某个列计算汇总,如查询 emp 表中每个部门的平均工资、工资总和等。

汇总函数也称分组函数,主要用于返回指定列的最大值、最小值、平均值、总和及非空列值数,函数名称分别为 max、min、avg、sum 及 count。

where 子句在分组之前过滤行,group by 子句指定分组列,having 子句过滤分组。

6.9.1 单独使用分组函数

以下是几个使用分组函数的示例。

查询 emp 表中 sal 列的最大、最小值。

```
mysql> select max(sal), min(sal) from emp;
```

查询 emp 表中 sal 列的总和及平均值。

```
mysql> select sum(sal), avg(sal) from emp;
```

查询 emp 表中 comm 列的非空值个数。

```
mysql> select count(comm) from emp;
```

count 函数的用法相对较多一点,在列名前面附加 distinct 关键字,可以查询非重复列值个数,参数改为 *,则可以查询表的总行数。

查询 emp 表的总行数,以及 deptno 列的非重复值个数。

```
mysql> select count(*), count(distinct deptno) from emp;
```

以上是分组函数的几个简单用法,使用分组函数时也可以附加限制条件。例如,查询 10 号部门的最高工资值。

```
mysql> select max(sal) from emp where deptno = 10;
```

6.9.2 使用 group by 子句执行分组汇总

group by 子句把指定列上有相同值的行分为一组,然后计算各组的汇总值。

在下面的示例中,group by 子句指定了分组列 deptno,其作用是把部门编号相同的行分为一组,即一个部门一组,然后计算每组的工资总和。

```
mysql> select deptno, sum(sal) from emp
    -> group by deptno;
+--------+----------+
| deptno | sum(sal) |
+--------+----------+
|     10 |  8750.00 |
|     20 |  6775.00 |
|     30 |  9400.00 |
+--------+----------+
3 rows in set (0.00 sec)
```

使用 group by 子句时,对语法有一些额外的要求。

select 子句中未被分组函数作用的列要在 group by 子句中选择,如下面的查询,在 select

子句中使用了 deptno,在 group by 子句中使用了 mgr,违反了此规则。

```
mysql> select deptno, sum(sal) from emp
    -> group by mgr;
ERROR 1055 (42000): Expression #1 of SELECT list is not in GROUP BY clause and contains nonaggregated column 'law.emp.deptno' which is not functionally dependent on columns in GROUP BY clause; this is incompatible with sql_mode = only_full_group_by
```

可以使用 where 子句在分组之前剔除不满足条件的行。

查询 emp 表中,每个部门工资超过 2 000 的员工的工资总和。

```
mysql> select deptno, sum(sal) from emp
    -> where sal > 2000
    -> group by deptno;
```

若使用 where 子句,在语法结构上要求 where 子句在 group by 子句之前。

要注意的是,因为 where 子句在分组之前就对记录进行了过滤,所以不能在 where 子句中使用分组函数,若要在查询条件中出现分组函数应使用 having 子句。

```
mysql> select deptno, sum(sal) from emp
    -> where sum(sal) > 7000
    -> group by deptno;
ERROR 1111 (HY000): Invalid use of group function
```

group by 子句可使用多个列名,按多个列分组,如查询每个部门中的工种个数,可以执行下面的命令。

```
mysql> select job,deptno,count(*) from emp
    -> group by job, deptno;
```

6.9.3 having 子句

having 子句用于过滤分组。如查询工资总额超过 7 000 的部门编号及其工资总额。

```
mysql> select deptno, sum(sal) from emp
    -> group by deptno
    -> having sum(sal) > 7000;
```

6.9.4 order by 子句

若要把分组汇总后的结果排序显示,可以使用 order by 子句。order by 子句总是放置在查询语句的最后。order by 子句中未被分组函数作用的列,只能从 group by 子句中选择。

以下查询中,同时出现了 where、group by、having 及 order by 子句,读者可以考查这几个子句的语法结构。

```
mysql> select deptno, sum(sal) from emp
    -> where sal > 2000
    -> group by deptno
    -> having avg(sal) > 1500
    -> order by sum(sal);
```

6.9.5 分组汇总查询小结

分组汇总查询可以使用多个子句,在语法结构上容易混淆,在此简单总结一下。

如果一个查询使用了前面小节所讲的各个子句,则 MySQL 按照以下顺序处理各子句:
- 执行 where 子句,过滤出符合条件的记录;
- 执行 group by 子句,对过滤后的结果按照指定列进行分组;
- 执行 having 子句,对分组进行过滤;
- 执行 order by 子句,对查询结果按照指定列排序;
- 执行 select 子句,显示最后结果。

对于语法规则,要注意以下几点:
- where 子句要放置在 group by 子句之前,不能在 where 子句中使用分组函数;
- having 子句放置在 group by 子句之后,一般会使用分组函数;
- 若未被分组函数作用,select、order by 子句中的列只能从 group by 子句选择。

6.10 子查询

子查询是一个 SQL 语句中包含的另外一个查询,这里主要说明查询语句中使用子查询。子查询所在的查询称为主查询。在语法结构方面,查询语句的各子句都可以使用子查询,下面分别举例说明。

6.10.1 where 子句中使用子查询

where 子句中使用子查询是比较典型的情况。如要求查询 emp 表中获得最高工资的员工名称。若用两个查询命令完成以上要求,则很简单。第 1 个查询得到最高工资值,第 2 个查询以此结果为条件查询员工名称,以下是查询过程。

```
mysql> select max(sal) from emp;
+----------+
| max(sal) |
+----------+
|  5000.00 |
+----------+
1 row in set (0.00 sec)

mysql> select ename from emp where sal = 5000;
+-------+
| ename |
+-------+
| KING  |
+-------+
1 row in set (0.00 sec)
```

若要求用一个命令完成以上要求,则可以把第 1 个查询作为子查询置于第 2 个查询的 where 子句中。

```
mysql> select ename from emp
    -> where sal = (select max(sal) from emp);
```

这样就克服了 where 子句不能使用分组函数的限制。

6.10.2 select 子句中使用子查询

在 select 子句中使用子查询,可以完成诸如下面的查询要求。

查询 emp 表中,SMITH 的工资与平均工资的差距。

```
mysql> select ename,sal - (select avg(sal) from emp) as "sal - avg(sal)"
    -> from emp where ename = 'SMITH';
+-------+----------------+
| ename | sal - avg(sal) |
+-------+----------------+
| SMITH |   -1277.083333 |
+-------+----------------+
1 row in set (0.00 sec)
```

如果在 select 子句中直接使用分组函数完成以上查询,则会出现语法错误:

```
mysql> select ename,sal - avg(sal) from emp where ename = 'SMITH';
ERROR 1140 (42000): In aggregated query without GROUP BY, expression #2 of SELECT list contains nonaggregated column 'law.emp.sal'; this is incompatible with sql_mode=only_full_group_by
```

6.10.3 from 子句中使用子查询

对表的查询结果可以看作另外一张表,这样,子查询可以像表一样放置在 from 子句中,但要注意对 from 子句中的子查询使用表别名,如下面的示例所示。

```
mysql> select * from
    ->(select deptno,avg(sal) from emp group by deptno) a;
+--------+-------------+
| deptno | avg(sal)    |
+--------+-------------+
|     10 | 2916.666667 |
|     20 | 2258.333333 |
|     30 | 1566.666667 |
+--------+-------------+
3 rows in set (0.01 sec)
```

上面的查询与只执行子查询的效果相同,看似多此一举,稍加修改后,可以克服 where 子句中不能使用分组函数的限制。

```
mysql> select * from
    ->(select deptno, avg(sal) as asal from emp group by deptno) a
    -> where asal > 2000;
```

6.10.4 非相关子查询与相关子查询

非相关子查询可以单独执行,不依赖主查询,示例参见 6.10.1 至 6.10.3 节内容。相关子查询的结果依赖主查询。相关子查询在逻辑结构上相对复杂,下面是几个示例。

查询 emp 表和 dept 表,得到每个员工的名称及其所属部门名称。

```
mysql> select ename,(select dname from dept where deptno = e.deptno) as dname
    -> from emp e;
+--------+------------+
| ename  | dname      |
+--------+------------+
| SMITH  | RESEARCH   |
| ALLEN  | SALES      |
...
| MILLER | ACCOUNTING |
+--------+------------+
12 rows in set (0.00 sec)
```

执行相关子查询,类似完成一个二重循环,对主查询的每一行,把子查询中出现的主查询列值代入后执行子查询,然后把主查询、子查询的结果显示出来。

查询 emp 表及 salgrade 表,得到每个员工名称及其工资级别。

```
mysql> select ename,
    -> (select grade from salgrade where e.sal between losal and hisal) as e
    -> from emp e;
+--------+-------+
| ename  | grade |
+--------+-------+
| SMITH  |   1   |
| ALLEN  |   3   |
...
| MILLER |   2   |
+--------+-------+
12 rows in set (0.00 sec)
```

查询 emp 表中每个员工名称及其经理名称。

```
mysql> select ename,(select ename from emp where empno = e.mgr)
    -> from emp e;
+--------+----------------------------------------------+
| ename  | (select ename from emp where empno = e.mgr)  |
+--------+----------------------------------------------+
| SMITH  | FORD                                         |
| ALLEN  | BLAKE                                        |
...                                                     |
| MILLER | CLARK                                        |
+--------+----------------------------------------------+
12 rows in set (0.00 sec)
```

查询 emp 表中工资超过其经理的员工名称及其经理名称。

```
mysql> select * from
    -> (
    ->     select ename,
    ->     (select ename from emp where empno = e.mgr and sal < e.sal) mgr_name
```

```
    ->     from emp e
    -> ) a
    -> where mgr_name is not null;
+-------+----------+
| ename | mgr_name |
+-------+----------+
| FORD  | JONES    |
+-------+----------+
1 row in set (0.00 sec)
```

上述查询要求也可以用下面的查询完成。

```
mysql> select ename,(select ename from emp where empno = e.mgr) as mgr_name
    -> from emp e
    -> where sal > (select sal from emp where empno = e.mgr);
```

查询 emp 表中,每个员工的工资与其所在部门的平均工资之差。

```
mysql> select ename,
    -> sal - (select avg(sal) from emp where deptno = e1.deptno ) as "sal - davg"
    -> from emp e1;
+--------+-------------+
| ename  | sal - davg  |
+--------+-------------+
| SMITH  | -1458.333333|
| ALLEN  |    33.333333|
...
| MILLER | -1616.666667|
+--------+-------------+
12 rows in set (0.00 sec)
```

查询每个部门的最高工资,要求在查询结果中列出部门名称。

```
mysql> select (select dname from dept where deptno = e.deptno) dname, max(sal) max_sal
    -> from emp e
    -> group by deptno;
+------------+---------+
| dname      | max_sal |
+------------+---------+
| RESEARCH   | 3000.00 |
| SALES      | 2850.00 |
| ACCOUNTING | 5000.00 |
+------------+---------+
3 rows in set (0.00 sec)
```

也可以执行下面的查询,但结果稍有不同。

```
mysql> select dname,
    -> (select max(sal) from emp where deptno = dept.deptno) max_sal
    -> from dept;
```

```
+------------+----------+
| dname      | max_sal  |
+------------+----------+
| ACCOUNTING | 5000.00  |
| RESEARCH   | 3000.00  |
| SALES      | 2850.00  |
| OPERATIONS |   NULL   |
+------------+----------+
4 rows in set (0.00 sec)
```

6.10.5　in 与 not in

in 和 not in 用于判断一个子查询结果中是否出现了某个表中的列值，有明显的元素与集合归属关系的意味，下面是几个典型示例。

如查询 dept 表中哪些部门还没有员工。

这里假设若 dept 表的某个 deptno 列值未在 emp 表中的 deptno 列出现，则认为这个部门没有员工。这样的话，这个查询要求也可以表述为：查询 dept 表中未在 emp 表中出现的部门编号对应的部门名称。

```
mysql> select dname from dept
    -> where deptno not in (select distinct deptno from emp);
+------------+
| dname      |
+------------+
| OPERATIONS |
+------------+
1 row in set (0.00 sec)
```

与上述命令的效果相反，如果要查询有员工的部门名称，只要把 not in 改为 in 即可。

```
mysql> select dname from dept
    -> where deptno in (select distinct deptno from emp);
+------------+
| dname      |
+------------+
| ACCOUNTING |
| RESEARCH   |
| SALES      |
+------------+
3 rows in set (0.00 sec)
```

对于 not in，有一个要注意的地方：当子查询的结果中包含 null 时，主查询不会有结果返回，下面的实验过程可以说明这一点。

先执行以下命令，把 ename 为 SMITH 的 deptno 值修改为空。

```
mysql> update emp set deptno = null where ename = 'SMITH';
```

下面命令的查询结果为空。

```
mysql> select dname from dept
    -> where deptno not in (select deptno from emp);
Empty set (0.00 sec)
```

这个结果可能会出乎一些读者的预料,简单解释一下出现这种情况的原因。

如果单独执行上面的子查询,结果中会存在一个空值。

```
mysql> select distinct deptno from emp;
+--------+
| deptno |
+--------+
|     20 |
|     30 |
|     10 |
|   NULL |
+--------+
4 rows in set (0.00 sec)
```

这样,原来的查询与下面的查询等价。

```
mysql> select dname from dept where deptno not in(null, 10, 20, 30);
Empty set (0.00 sec)
```

而这个查询与下面的查询又是等价的。

```
mysql> select dname from dept
    -> where deptno != null and deptno != 10 and deptno != 20 and deptno != 30;
Empty set (0.00 sec)
```

关系表达式 deptno != null 中包含了 null,其布尔值为假,不论其后的 3 个关系表达式的结果是真是假,经过且(即 and)运算后,where 子句中的整个逻辑表达式的结果总为假,因而当子查询中含有 null 值时,主查询结果总为空。

为了解决此问题,使用 not in 时,可以在子查询中附加"is not null"条件过滤 null 值,如下面命令所示。

```
mysql> select dname from dept
    -> where deptno not in(select distinct deptno from emp where deptno is not null)
```

如果使用 in 查询有员工的部门,即使 deptno 列存在 null 值也会得到正确结果。此时的条件经历了如下变化:

deptno in (select deptno from emp)→

deptno in(null,10,20,30)→

deptno=null or deptno=10 or deptno=20 or deptno=30

dept 表中 deptno 列除了 40 以外的 3 个值均可以使最后一个逻辑表达式中的 4 个判断条件中的某一个为真,从而使得整个判断结果为真。

这里 in 和 not in 的用法也分别称为半连接(semi-join)和反连接(anti-join)。

上面 in 的用法还有另外一种等价的方法,用关键字 any 来实现,如

```
mysql> select dname from dept where deptno in(select distinct deptno from emp);
```

可以改写为

```
mysql> select dname from dept where deptno = any(select distinct deptno from emp);
```

同样,下面 not in 语句

```
mysql> select dname from dept
    -> where deptno not in(select distinct deptno from emp where deptno is not null);
```

其等价用法可以使用关键字 all 实现,

```
mysql> select dname from dept
    -> where deptno <> all(select distinct deptno from emp where deptno is not null);
```

6.10.6　exists 与 not exists

exists 和 not exists 的用法与 in 和 not in 相似,区别是这两个子句一般使用相关子查询,如查询 dept 表中有员工和无员工的部门名称,可以使用 exists 和 not exists 分别实现。

使用 exists 查询有员工的部门名称。

```
mysql> select dname from dept
    -> where exists(select deptno from emp where deptno = dept.deptno);
```

使用 not exists 查询没有员工的部门名称。

```
mysql> select dname from dept
    -> where not exists(select deptno from emp where deptno = dept.deptno);
```

对于 dept 表中每行的 deptno 列值,exists 测试此值是否使其后的子查询结果为空集,若非空集,则返回此 deptno 列值对应的 dname 值。not exists 与此相反,若此值使其后的子查询结果为空集,则返回此 deptno 列值对应的 dname 值。

与 not in 不同,当子查询的返回结果中存在 null 值时,使用 not exists 依然可以返回满足条件的结果,而不是空集。

检查 emp 表的 deptno 列是否存在 null 值,若不存在,则执行以下命令设置 ename 为 SMITH 的 deptno 为 null。

```
mysql> update emp set deptno = null where ename = 'SMITH';
```

执行上述修改操作后,emp 表的 deptno 列值为 null、10、20、30,重新使用 not exists 子句执行查询。

```
mysql> select dname from dept
    -> where not exists(select deptno from emp where deptno = dept.deptno);
+------------+
| dname      |
+------------+
| OPERATIONS |
+------------+
1 row in set (0.00 sec)
```

可以发现,不同于使用 not in 的情况,emp 表的 deptno 列上的 null 值并未导致以上查询结果为空。

当主查询中 dept 表的 deptno 列值遍历至 10、20 或 30 中的任何一个时,即使 emp 表的 deptno 列值有 null 值,子查询也会分别返回 10、20 或 30,即子查询的返回结果不为空集,从而这 3 个 deptno 值对应的 3 个 dname 不会作为主查询的返回结果。当主查询 deptno 列值遍历至 40 时,emp 表的 4 个 deptno 列值(即 null,10,20,30)中的任何一个,都会使子查询结果为空集,因此与 deptno = 40 对应的 dname,即 OPERATIONS 总会返回。

6.11 集合运算

SQL 语言的集合运算用来实现数学中的集合运算。SQL 标准中的集合运算包括 union、union all、intersect 及 except，前两种求并集，后两种分别求交集和差集。MySQL 当前只支持 union 和 union all。

union 把两个查询的结果合并，去掉重复行后返回结果。union all 只是返回合并结果，不去掉重复行。

例如，返回 ename 首字母为 A 或尾字母为 B 的记录，并消除重复行。

```
mysql> select ename,deptno,sal from emp where ename like 'A%'
    -> union
    -> select ename,deptno,sal from emp where ename like '%B';
```

6.12 多表连接查询

当查询的数据存在于多个表时，需要用到表连接。如查询每个员工的名称及其所在部门名称，要涉及 emp 和 dept 两个表。表连接一般可以分为交叉连接、内连接与外连接，下面分别以实例说明。

6.12.1 交叉连接

交叉连接是对两个表执行不附加连接条件的查询方式。

下面示例对 emp 和 dept 表执行交叉连接查询，为了让显示效果清晰，emp 表取出 ename 及 deptno 列，dept 表取出所有列。

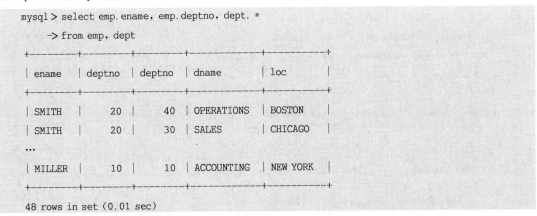

分析上面查询结果，不难得出以下结论：如果把以上交叉连接查询结果看成另外一个表，则这个表的记录是由两个表中的行两两横向拼接得到的。emp 表取出 2 列，dept 表取出 3 列，结果表由 5 列构成，emp 表有 12 行记录，dept 表有 4 行记录，查询结果表由 48（即 12×4）行记录构成。

这些结论具有一般性：对两个表做交叉连接时，查询结果可以看作一个新表。这个表的列由两个表的列合并得到，如一个表 5 个列，另一个表 3 个列，查询结果由 8 个列组成；这个表的记录由第 1 个表的每行记录与第 2 个表的每行记录横向拼接而成，如果第 1 个表有 m 行记录,第 2 个表有 n 行记录,则查询结果由 $(m \times n)$ 行记录构成。

6.12.2 内连接

在交叉连接的查询结果中,多数都是没有意义的拼接结果,如果要从中获得有意义的结果,则需要附加连接条件。交叉连接附加连接条件后,称为内连接,其查询结果由两个表中满足连接条件的记录拼接而成。

如在 6.12.1 的交叉连接查询结果中,两个 deptno 列若列值相等,则可以看作有意义的拼接结果:前面是员工名称,后面是其所属部门的信息。得到这些有意义的结果,可以附加连接条件:emp.deptno = dept.deptno,如下面查询所示。

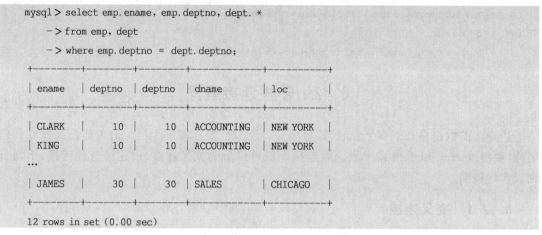

为了可读性更好,在连接查询中可以使用表别名引用各自的列,以上命令可修改为

```
mysql> select e.ename, e.deptno, d.* from emp e, dept d where e.deptno = d.deptno;
```

连接条件一般用等式表示,有时也会用到不等式,如查询每个员工的名称及其工资级别,需要在 emp 表和 salgrade 表的交叉连接结果中,使用不等式过滤出有意义的拼接数据。

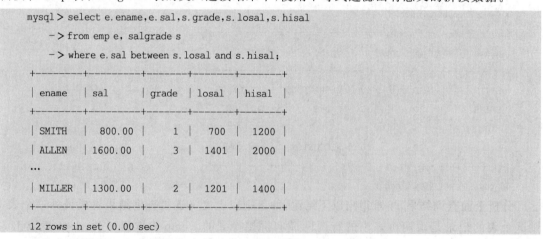

有些查询需要一个表与其自身做连接,即自连接查询。如查询每个员工的名称及其经理名称,需要 emp 表与其自身连接。注意到 emp 表中的 mgr 列值表示其所在行员工的经理员工号,可以构造出以下命令。

```
mysql> select e.ename,e.mgr,m.empno,m.ename
    -> from emp e, emp m
    -> where e.mgr = m.empno;
+--------+------+-------+--------+
| ename  | mgr  | empno | ename  |
+--------+------+-------+--------+
| SMITH  | 7902 | 7902  | FORD   |
| ALLEN  | 7698 | 7698  | BLAKE  |
...
| MILLER | 7782 | 7782  | CLARK  |
+--------+------+-------+--------+
11 rows in set (0.00 sec)
```

若只需要员工名称及其经理名称,则可以在 select 子句中去除不必要的 mgr 和 empno 列,修改为如下形式。

```
mysql> select e.ename,m.ename
    -> from emp e, emp m
    -> where e.mgr = m.empno;
```

查询工资超过其经理的员工的名称,只需要减少 select 子句中的列,并增加一个条件:

```
mysql> select e.ename
    -> from emp e, emp m
    -> where e.mgr = m.empno and e.sal > m.sal;
```

很多情况下,需要执行 3 个表,甚至 3 个表以上的连接查询,这时需要把参与连接查询的多个表放置在 from 子句中,并增加相应连接条件。

如查询每个员工名称、所在部门名称及其工资级别。

```
mysql> select e.ename,d.dname,s.grade
    -> from emp e, dept d, salgrade s
    -> where e.deptno = d.deptno and e.sal between s.losal and s.hisal;
+--------+------------+-------+
| ename  | dname      | grade |
+--------+------------+-------+
| SMITH  | RESEARCH   |   1   |
| ALLEN  | SALES      |   3   |
...
| MILLER | ACCOUNTING |   2   |
+--------+------------+-------+
12 rows in set (0.00 sec)
```

仿照以上语法形式可以很容易地构造出 4 个表的连接查询,这里不再赘述。

6.12.3　两种连接标准:SQL-86 与 SQL-92

表连接查询有两种写法,分别符合 SQL-86 和 SQL-92 标准,上述内容涉及的语法形式符合 SQL-86 标准,其特点是在 from 子句中放置多个表名,在 where 子句中放置连接条件。

而 SQL-92 标准使用 inner join 关键字连接两个表(inner 可以省略),涉及两个表的连接条件放置于 on 子句。只涉及单个表的限制条件虽然也可以放置于 on 子句,但习惯上放置于 where 子句。对于交叉连接,SQL-92 标准使用 cross join 关键字。下面通过实例说明 SQL-92 标准的语法形式。

使用 SQL-92 语法形式,对 emp 表和 dept 表执行交叉连接。

```
mysql> select * from emp cross join dept;
```

使用 SQL-92 语法的内连接形式,查询每个员工的名称及其所在部门名称:

```
mysql> select e.ename,d.dname
    -> from emp e join dept d on e.deptno = d.deptno;
```

涉及单个表的条件可以放置于 on 子句,也可以放置于 where 子句,如查询 10 部门中的每个员工名称及其所在部门名称,以下两种语法形式都符合语法规则,效果也相同,但最好使用后一种形式。

```
mysql> select e.ename,d.dname
    -> from emp e join dept d on e.deptno = d.deptno and e.deptno = 10;

mysql> select e.ename,d.dname
    -> from emp e join dept d on e.deptno = d.deptno
    -> where e.deptno = 10;
```

对于 3 个表或 3 个表以上的连接查询,可以先执行前两个表的连接,再与第 3 个表连接,如查询每个员工的名称、部门名称及其工资级别。

```
mysql> select e.ename, d.dname, s.grade
    -> from emp e join dept d on e.deptno = d.deptno
    -> join salgrade s on e.sal between s.losal and s.hisal;
```

SQL-86 和 SQL-92 两种连接标准没有优劣之分,可以根据自己的喜好选择其一。

6.12.4 外连接

内连接的显示结果是两表中符合连接条件的数据,不满足连接条件的数据则未显示,如果需要把两者都显示出来,可以使用外连接。

外连接分为左外连接、右外连接及全外连接,SQL 标准分别使用 left outer join,right outer join 及 full outer join 表示,MySQL 尚不支持全外连接。

3 种外连接先把符合内连接的数据显示出来,另外的区别如下。

- 左外连接:也把"left outer join"左侧表中不满足连接条件的数据显示出来。
- 右外连接:也把"right outer join"右侧表中不满足连接条件的数据显示出来。
- 全外连接:也把"full outer join"两侧表中不满足连接条件的数据显示出来。

下面用几个示例,分别说明几种外连接的用法。

先对 emp 表添加一行新纪录,使其 deptno 为空。

```
mysql> insert into emp(empno,ename) values(8888,'MIKE');
```

使用左外连接查询每个员工的名称及其所在部门的名称。

```
mysql> select e.ename,d.dname from emp e left outer join dept d
    -> on e.deptno = d.deptno;
```

```
+--------+------------+
| ename  | dname      |
+--------+------------+
| SMITH  | RESEARCH   |
| ALLEN  | SALES      |
...
| MILLER | ACCOUNTING |
| MIKE   | NULL       |
+--------+------------+
13 rows in set (0.00 sec)
```

可以发现,符合连接条件的内连接数据显示在前面,不符合连接条件的左侧表数据显示在最后,其对应右侧表的数据为空。

使用右外连接查询每个员工的名称及其所在部门的名称。

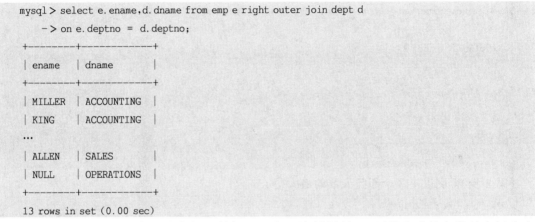

与左外连接相似,符合连接条件的数据显示在查询结果前面,不符合连接条件的右侧表中的数据显示在最后,其对应左侧表的数据为空。

MySQL 不支持全外连接,可以使用 union 间接实现,如查询每个员工的名称及其所在部门的名称。

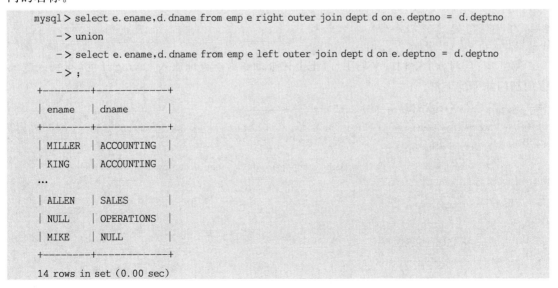

在进行下面的实验之前,请删除本节对 emp 表添加的记录。

```
mysql> delete from emp where empno = 8888;
```

6.13 构造复杂的查询语句

如果没有可遵循的步骤,要构造一个复杂的 SQL 查询,往往很困难。本节以一个实例说明构造复杂查询语句的一般步骤。

本节实例的查询要求是:得出人数最多的部门的名称及该部门平均工资的级别。

得到部门平均工资,需要查询 emp 表;得到平均工资的级别,需要查询 salgrade 表;得到部门名称,需要查询 dept 表,完成此查询要使用 3 个表的连接。下面由简单到复杂逐步构造出这个查询语句。

第 1 步:查询 emp 表中每个部门的人数。

```
mysql> select deptno, count(*) from emp group by deptno;
+--------+----------+
| deptno | count(*) |
+--------+----------+
|     20 |        3 |
|     30 |        6 |
|     10 |        3 |
+--------+----------+
3 rows in set (0.00 sec)
```

第 2 步:得到以上 3 个部门人数的最大值。

```
mysql> select max(cnt)
    -> from (select count(*) cnt from emp group by deptno) a;
+----------+
| max(cnt) |
+----------+
|        6 |
+----------+
1 row in set (0.00 sec)
```

第 3 步:查询每个部门的平均工资,把上述查询作为子查询放入 having 子句,得到人数最多的部门的平均工资。

```
mysql> select deptno, avg(sal) from emp
    -> group by deptno
    -> having count(*) =
    -> (select max(cnt) from (select count(*) cnt from emp group by deptno) a);
+--------+-------------+
| deptno | avg(sal)    |
+--------+-------------+
|     30 | 1566.666667 |
+--------+-------------+
1 row in set (0.00 sec)
```

第4步：把以上语句作为子查询放入 from 子句，与 salgrade 表、dept 表做交叉连接。

```
mysql> select d.deptno,d.dname,dm.deptno,dm.avg_sal,s.grade,s.losal,s.hisal
    -> from dept d,
    -> (select deptno,avg(sal) avg_sal from emp group by deptno
    -> having count(*) =
    -> (select max(cnt) from (select count(*) cnt from emp group by deptno) a)) dm,
    -> salgrade s;
+--------+------------+--------+-------------+-------+-------+-------+
| deptno | dname      | deptno | avg_sal     | grade | losal | hisal |
+--------+------------+--------+-------------+-------+-------+-------+
|     10 | ACCOUNTING |     30 | 1566.666667 |     5 |  3001 |  9999 |
|     20 | RESEARCH   |     30 | 1566.666667 |     5 |  3001 |  9999 |
...
|     40 | OPERATIONS |     30 | 1566.666667 |     1 |   700 |  1200 |
+--------+------------+--------+-------------+-------+-------+-------+
20 rows in set (0.00 sec)
```

第5步：分析查询结果，构造连接条件，过滤出有意义的行。经过分析，3个表需要有两个连接条件。

第1个连接条件是前两个表的 deptno 相等。

第2个连接条件是平均工资值应该在 salgrade 表的 losal 和 hisal 之间。

由此可以得出以下查询命令。

```
mysql> select d.deptno, d.dname, dm.deptno, dm.avg_sal, s.grade, s.losal, s.hisal
    -> from dept d,
    -> (
    ->    select deptno,avg(sal) avg_sal
    ->    from emp
    ->    group by deptno
    ->    having count(*) =
    ->      (select max(cnt) from (select count(*) cnt from emp group by deptno) a)
    -> ) dm,
    -> salgrade s
    -> where d.deptno = dm.deptno
    -> and dm.avg_sal between s.losal and s.hisal;
+--------+-------+--------+-------------+-------+-------+-------+
| deptno | dname | deptno | avg_sal     | grade | losal | hisal |
+--------+-------+--------+-------------+-------+-------+-------+
|     30 | SALES |     30 | 1566.666667 |     3 |  1401 |  2000 |
+--------+-------+--------+-------------+-------+-------+-------+
1 row in set (0.00 sec)
```

第6步：在 select 子句中剔除不必要的列，得出最终查询结果。

```
mysql> select d.deptno, d.dname, dm.avg_sal, s.grade
    -> from dept d,
    -> (
```

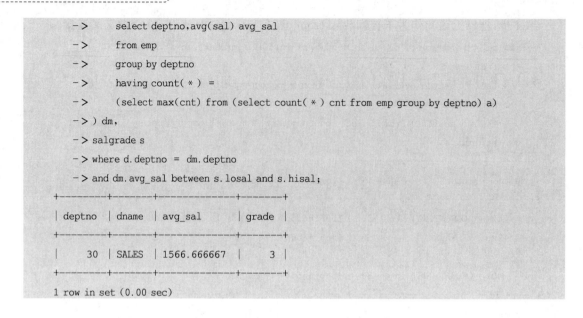

6.14 SQL 查询的等效转换

一个查询要求,一般存在多种实现方式,不同的方式有时会有不同的效率。本节把效果等价的不同查询放在一起进行对比。

6.14.1 内连接与子查询

查询 emp 表和 dept 表得出每个员工的名称及其部门名称是典型的内连接查询,也可以用子查询完成,为了对比,下面把这两个效果等价的查询罗列在一起。

使用内连接的形式如下。

```
mysql> select e.ename,d.dname from emp e,dept d
    -> where e.deptno = d.deptno;
```

使用子查询的形式如下。

```
mysql> select ename,(select dname from dept where deptno = e.deptno) as dname
    -> from emp e;
```

6.14.2 in,exists,内连接

从前面的内容可知,in 和 exists 可以相互转换,如查询 dept 表中有员工的部门名称,以下两种方式都是可以的。

使用 in 的方式:

```
mysql> select dname from dept
    -> where deptno in(select deptno from emp);
```

使用 exists 的方式:

```
mysql> select dname from dept
    -> where exists(select deptno from emp where deptno = dept.deptno);
```

如果使用内连接,以上查询要求也可以用以下方式:

```
mysql> select distinct d.dname from emp e,dept d where e.deptno = d.deptno;
```

还可以使用 any 关键字实现：

```
mysql> select dname from dept where deptno = any(select distinct deptno from emp);
```

6.14.3　not in,not exist,外连接

in 和 exists 可以相互转换，not in 和 not exists 也可以相互转换，如查询 dept 表中没有员工的部门名称，以下两种方式都可以实现。

使用 not in 的方式：

```
mysql> select dname from dept
    -> where deptno not in(select deptno from emp where deptno is not null);
```

使用 not exists 的方式：

```
mysql> select dname from dept
    -> where not exists(select deptno from emp where deptno = dept.deptno);
```

如果使用外连接，则以上查询要求也可以用以下方式：

```
mysql> select d.dname from emp e right outer join dept d on e.deptno = d.deptno
    -> where e.deptno is null;
```

还可以用以下方式：

```
mysql> select dname from dept
    -> where deptno <> all(select distinct deptno from emp where deptno is not null);
```

第 7 章 窗口函数

在数据库中执行统计操作一般使用汇总函数结合 group by 子句,这种方式在语法上有一定的限制,窗口函数可以避开这些限制,完成更强的统计功能。窗口函数已加入 SQL-2003 中,主流数据库产品都已经对其提供支持。

本章主要内容包括：
- over()子句构造窗口；
- 窗口内划分框架；
- 窗口函数分类与示例。

7.1 over()子句构造窗口

使用窗口函数时,需要附加 over()子句构造窗口,与当前行相关的若干行构成一个窗口。

执行普通的分组汇总,先由 group by 子句产生分组,然后对每个组得出一个统计结果,查询结果的行数与组数相同。使用窗口函数执行统计操作时,每个行都有与其相关的窗口,每个行会得到一个窗口统计结果。

over()子句可以为空,也可以包含下面的子句。
- partition by:指定列名,划分窗口。
- order by:指定列名,对窗口内的行进行排序。
- frame:在窗口内,以当前行为基准,由其之前或之后范围内的若干行构成的子窗口,一般译为框架。划分框架后,统计操作以框架为单位进行。

下面以求和函数 sum()为例,简单说明以上各个子句的功能及用法。

7.1.1 partition by 子句构造窗口

若 over 子句为空,则窗口视为整个表。下面示例中的 over 子句为空,sum(sal)以整个表为窗口计算 sal 之和,win_sum 为其列别名。

```
mysql> select ename, deptno, sal, sum(sal) over() win_sum from emp;
+--------+--------+---------+----------+
| ename  | deptno | sal     | win_sum  |
+--------+--------+---------+----------+
| SMITH  |     20 |  800.00 | 24925.00 |
| ALLEN  |     30 | 1600.00 | 24925.00 |
...
| MILLER |     10 | 1300.00 | 24925.00 |
+--------+--------+---------+----------+
12 rows in set (0.00 sec)
```

如果附加了 where 子句,则先过滤出符合条件的行,窗口由符合条件的所有行构成。

```
mysql> select ename, deptno, sal, sum(sal) over() win_sum from emp
    -> where deptno = 10;
+--------+--------+---------+---------+
| ename  | deptno | sal     | win_sum |
+--------+--------+---------+---------+
| CLARK  |     10 | 2450.00 | 8750.00 |
| KING   |     10 | 5000.00 | 8750.00 |
| MILLER |     10 | 1300.00 | 8750.00 |
+--------+--------+---------+---------+
3 rows in set (0.01 sec)
```

在 over() 子句中使用 partition by deptno 子句创建窗口,deptno 相同的行会各自构成一个窗口,sum(sal)为每个窗口内的 sal 之和。

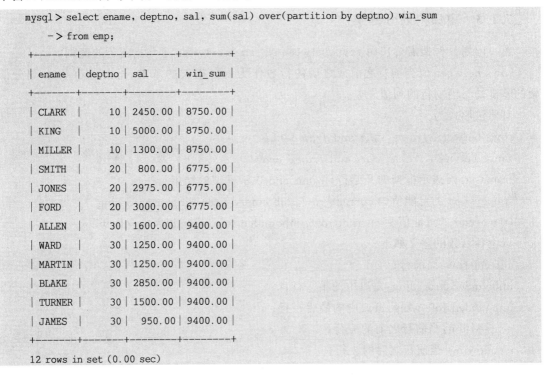

7.1.2 order by 子句设置窗口内排序

若再附加 order by ename 子句,sum(sal)会在每个窗口内按 ename 列值顺序统计累加和。

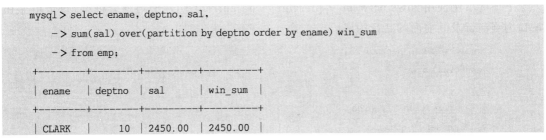

```
|   KING   |  10 | 5000.00 | 7450.00 |
|  MILLER  |  10 | 1300.00 | 8750.00 |
|   FORD   |  20 | 3000.00 | 3000.00 |
|  JONES   |  20 | 2975.00 | 5975.00 |
|  SMITH   |  20 |  800.00 | 6775.00 |
|  ALLEN   |  30 | 1600.00 | 1600.00 |
|  BLAKE   |  30 | 2850.00 | 4450.00 |
|  JAMES   |  30 |  950.00 | 5400.00 |
|  MARTIN  |  30 | 1250.00 | 6650.00 |
|  TURNER  |  30 | 1500.00 | 8150.00 |
|   WARD   |  30 | 1250.00 | 9400.00 |
+----------+-----+---------+---------+
12 rows in set (0.00 sec)
```

7.1.3 窗口内划分框架

窗口内划分框架需要使用 rows between 或 range between 子句指定行的范围。

rows between 以当前行之前或之后的行数作为划分框架的基准；range between 划分框架的基准是与当前行的列值之差。

其语法形式为

rows between *frame_start* and *frame_end*

range between *frame_start* and *frame_end*

frame_start 表示框架的开始行，frame_end 表示框架的结束行。

frame_start 的可能值为 current_row、unbounded preceding、*n* preceding。

frame_end 的可能值为 current row、unbounded following 以及 *n* following。

以上各参数的意义如下。

current row：当前行；

unbounded preceding：窗口中的第一行；

unbounded following：窗口中的最后一行；

n preceding：当前行之前的 *n* 行；

n following：当前行之后的 *n* 行。

preceding 和 following 也可以使用日期时间差作为划分标准，如：

interval 5 day preceding；

interval '2:30' minute_second following。

下面通过几个示例说明各个参数的用法。

下面的示例使用 rows between 2 preceding and current row 子句划分框架，每个框架由窗口内当前行及其前面的 2 行构成。

```
mysql> select ename, deptno, sal,
    -> sum(sal) over
    -> (
    ->        partition by deptno
    ->        order by ename
```

```
    ->         rows between 2 preceding and current row
    -> ) win_sum
    -> from emp;
+--------+--------+---------+---------+
| ename  | deptno | sal     | win_sum |
+--------+--------+---------+---------+
| CLARK  |     10 | 2450.00 | 2450.00 |
| KING   |     10 | 5000.00 | 7450.00 |
| MILLER |     10 | 1300.00 | 8750.00 |
| FORD   |     20 | 3000.00 | 3000.00 |
| JONES  |     20 | 2975.00 | 5975.00 |
| SMITH  |     20 |  800.00 | 6775.00 |
| ALLEN  |     30 | 1600.00 | 1600.00 |
| BLAKE  |     30 | 2850.00 | 4450.00 |
| JAMES  |     30 |  950.00 | 5400.00 |
| MARTIN |     30 | 1250.00 | 5050.00 |
| TURNER |     30 | 1500.00 | 3700.00 |
| WARD   |     30 | 1250.00 | 4000.00 |
+--------+--------+---------+---------+
12 rows in set (0.00 sec)
```

下面示例的框架由窗口内当前行及其前面的所有行构成。

```
mysql> select ename, deptno, sal, sum(sal) over
    -> (
    ->     partition by deptno
    ->     order by ename
    ->     rows between unbounded preceding and current row
    -> ) win_sum
    -> from emp;
+--------+--------+---------+---------+
| ename  | deptno | sal     | win_sum |
+--------+--------+---------+---------+
| CLARK  |     10 | 2450.00 | 2450.00 |
| KING   |     10 | 5000.00 | 7450.00 |
| MILLER |     10 | 1300.00 | 8750.00 |
| FORD   |     20 | 3000.00 | 3000.00 |
| JONES  |     20 | 2975.00 | 5975.00 |
| SMITH  |     20 |  800.00 | 6775.00 |
| ALLEN  |     30 | 1600.00 | 1600.00 |
| BLAKE  |     30 | 2850.00 | 4450.00 |
| JAMES  |     30 |  950.00 | 5400.00 |
| MARTIN |     30 | 1250.00 | 6650.00 |
| TURNER |     30 | 1500.00 | 8150.00 |
| WARD   |     30 | 1250.00 | 9400.00 |
+--------+--------+---------+---------+
12 rows in set (0.00 sec)
```

使用 range 形式,框架范围是当前行及其之前 sal 差值在 2 000 以内的行。

```
mysql> select ename, deptno, sal, sum(sal) over
    -> (
    ->     partition by deptno
    ->     order by sal
    ->     range between 2000 preceding and current row
    -> ) win_sum
    -> from emp;
+--------+--------+---------+---------+
| ename  | deptno | sal     | win_sum |
+--------+--------+---------+---------+
| MILLER |     10 | 1300.00 | 1300.00 |
| CLARK  |     10 | 2450.00 | 3750.00 |
| KING   |     10 | 5000.00 | 5000.00 |
| SMITH  |     20 |  800.00 |  800.00 |
| JONES  |     20 | 2975.00 | 2975.00 |
| FORD   |     20 | 3000.00 | 5975.00 |
| JAMES  |     30 |  950.00 |  950.00 |
| WARD   |     30 | 1250.00 | 3450.00 |
| MARTIN |     30 | 1250.00 | 3450.00 |
| TURNER |     30 | 1500.00 | 4950.00 |
| ALLEN  |     30 | 1600.00 | 6550.00 |
| BLAKE  |     30 | 2850.00 | 9400.00 |
+--------+--------+---------+---------+
12 rows in set (0.00 sec)
```

下面示例使用 hiredate 列的日期差作为划分框架的基准，框架由当前行及其之前 3 个月内的行构成。

```
mysql> select ename, deptno, sal, hiredate, sum(sal) over
    -> (
    ->     partition by deptno
    ->     order by hiredate
    ->     range between interval 3 month preceding and current row
    -> ) win_sum
    -> from emp;
+--------+--------+---------+------------+---------+
| ename  | deptno | sal     | hiredate   | win_sum |
+--------+--------+---------+------------+---------+
| CLARK  |     10 | 2450.00 | 1981-06-09 | 2450.00 |
| KING   |     10 | 5000.00 | 1981-11-17 | 5000.00 |
| MILLER |     10 | 1300.00 | 1982-01-23 | 6300.00 |
| SMITH  |     20 |  800.00 | 1980-12-17 |  800.00 |
| JONES  |     20 | 2975.00 | 1981-04-02 | 2975.00 |
| FORD   |     20 | 3000.00 | 1981-12-03 | 3000.00 |
| ALLEN  |     30 | 1600.00 | 1981-02-02 | 1600.00 |
| WARD   |     30 | 1250.00 | 1981-02-22 | 2850.00 |
| BLAKE  |     30 | 2850.00 | 1981-05-01 | 5700.00 |
| TURNER |     30 | 1500.00 | 1981-09-08 | 1500.00 |
| MARTIN |     30 | 1250.00 | 1981-09-28 | 2750.00 |
| JAMES  |     30 |  950.00 | 1981-12-03 | 3700.00 |
+--------+--------+---------+------------+---------+
12 rows in set (0.00 sec)
```

7.1.4 命名窗口

如果一个查询多次调用窗口函数,而窗口为同一个,则可以使用命名窗口简化其写法。命名窗口使用 window 关键字。

如下面语句,

```
mysql> select ename, deptno, sal,
    -> sum(sal) over(partition by deptno order by ename) win_sum
    -> from emp;
```

可以改写为

```
mysql> select ename, deptno, sal, sum(sal) over w win_sum from emp
    -> window w as (partition by deptno order by ename);
```

在查询语句的多个窗口函数中使用命名窗口:

```
mysql> select ename, deptno, sal,
    -> sum(sal) over w win_sum,
    -> avg(sal) over w win_avg
    -> from emp
    -> window w as (partition by deptno order by ename);
```

7.2 窗口函数分类与示例

窗口函数总体上分为汇总函数和排名函数。这里的汇总函数名称与简单查询时使用的汇总函数相同。如果在使用汇总函数时,也使用了 over() 子句,则这个汇总函数作为窗口函数使用。

7.2.1 汇总函数

下面的示例说明几个常用的汇总函数作为窗口函数的用法。

```
mysql> select ename, deptno,
    -> sum(sal) over w "sum_sal",
    -> avg(sal) over w "avg_sal",
    -> max(sal) over w "max_sal",
    -> min(sal) over w "min_sal",
    -> count(*) over w "cnt"
    -> from emp
    -> window w as (partition by deptno);
```

ename	deptno	sum_sal	avg_sal	max_sal	min_sal	cnt
CLARK	10	8750.00	2916.666667	5000.00	1300.00	3
KING	10	8750.00	2916.666667	5000.00	1300.00	3
MILLER	10	8750.00	2916.666667	5000.00	1300.00	3
SMITH	20	6775.00	2258.333333	3000.00	800.00	3
JONES	20	6775.00	2258.333333	3000.00	800.00	3

```
| FORD    |   20 | 6775.00 | 2258.333333 | 3000.00 |  800.00 | 3 |
| ALLEN   |   30 | 9400.00 | 1566.666667 | 2850.00 |  950.00 | 6 |
| WARD    |   30 | 9400.00 | 1566.666667 | 2850.00 |  950.00 | 6 |
| MARTIN  |   30 | 9400.00 | 1566.666667 | 2850.00 |  950.00 | 6 |
| BLAKE   |   30 | 9400.00 | 1566.666667 | 2850.00 |  950.00 | 6 |
| TURNER  |   30 | 9400.00 | 1566.666667 | 2850.00 |  950.00 | 6 |
| JAMES   |   30 | 9400.00 | 1566.666667 | 2850.00 |  950.00 | 6 |
+---------+------+---------+-------------+---------+---------+---+
12 rows in set (0.00 sec)
```

7.2.2 排名函数

排名函数根据功能分为两类:获得总体排名和获得指定名次。

获得总体排名函数包括下面 6 个。

- row_number():对查询结果中的行附加行号,行号不重复。
- rank():对窗口内的行排名,出现相同名次时,下一个名次会跳跃。
- dense_rank():对窗口内的行排名,出现相同名次时,下一个名次不会跳跃。
- percent_rank():返回(当前行位次－1)/(总行数－1),对于重复值,取第一个行的位置。
- cume_dist():返回(当前行位次)/(总行数),对于重复值,取最后一个行的位置。
- ntile(n):对窗口内的行根据排序列依次分为 n 个行数均等的组,并从 1 开始,由小到大赋予每个组一个编号。

以上函数除 row_number()外,均要在 over()子句中使用 order by,否则各行的排名相同。

获得指定名次的函数包括下面 5 个(参数 *col* 为列名)。

- first_value(*col*):返回窗口的第一行的 *col* 列值。
- last_value(*col*):返回当前行的 *col* 列值,如果要返回窗口的最后一行的 *col* 列值,则需要使用框架包括整个窗口。
- nth_value(*col*, *m*):返回当前窗口的第 *m* 行。
- lag(*col*, *n*):返回当前行之前的第 *n* 行。
- lead(*col*, *n*):返回当前行之后的第 *n* 行。

下面的示例以整个 emp 表为一个窗口,按照 sal 值得到其各种排名。

```
mysql> select ename, sal,
    -> row_number() over w "row_num",
    -> rank() over w "rank",
    -> dense_rank() over w "den_rank",
    -> ntile(4) over w "ntile"
    -> from emp
    -> window w as (order by sal);
+-------+--------+---------+------+----------+-------+
| ename | sal    | row_num | rank | den_rank | ntile |
+-------+--------+---------+------+----------+-------+
| SMITH | 800.00 |       1 |    1 |        1 |     1 |
```

```
| JAMES  |  950.00 |   2 |   2 |   2 |   1 |
| MARTIN | 1250.00 |   3 |   3 |   3 |   1 |
| WARD   | 1250.00 |   4 |   3 |   3 |   2 |
| MILLER | 1300.00 |   5 |   5 |   4 |   2 |
| TURNER | 1500.00 |   6 |   6 |   5 |   2 |
| ALLEN  | 1600.00 |   7 |   7 |   6 |   3 |
| CLARK  | 2450.00 |   8 |   8 |   7 |   3 |
| BLAKE  | 2850.00 |   9 |   9 |   8 |   3 |
| JONES  | 2975.00 |  10 |  10 |   9 |   4 |
| FORD   | 3000.00 |  11 |  11 |  10 |   4 |
| KING   | 5000.00 |  12 |  12 |  11 |   4 |
+--------+---------+-----+-----+-----+-----+
12 rows in set (0.00 sec)
```

以下示例查看各行的两种百分位数。

```
mysql> select ename, sal,
    -> percent_rank() over w "per_rank", cume_dist() over w "cume_dist"
    -> from emp
    -> window w as (order by sal);
+--------+---------+---------------------+---------------------+
| ename  | sal     | per_rank            | cume_dist           |
+--------+---------+---------------------+---------------------+
| SMITH  |  800.00 |                   0 | 0.08333333333333333 |
| JAMES  |  950.00 | 0.09090909090909091 | 0.16666666666666666 |
| MARTIN | 1250.00 | 0.18181818181818182 | 0.3333333333333333  |
| WARD   | 1250.00 | 0.18181818181818182 | 0.3333333333333333  |
| MILLER | 1300.00 | 0.36363636363636365 | 0.4166666666666667  |
| TURNER | 1500.00 | 0.45454545454545453 |                 0.5 |
| ALLEN  | 1600.00 | 0.5454545454545454  | 0.5833333333333334  |
| CLARK  | 2450.00 | 0.6363636363636364  | 0.6666666666666666  |
| BLAKE  | 2850.00 | 0.7272727272727273  |                0.75 |
| JONES  | 2975.00 | 0.8181818181818182  | 0.8333333333333334  |
| FORD   | 3000.00 | 0.9090909090909091  | 0.9166666666666666  |
| KING   | 5000.00 |                   1 |                   1 |
+--------+---------+---------------------+---------------------+
12 rows in set (0.00 sec)
```

以下示例查询每个窗口指定行的 sal 值。

```
mysql> select ename, sal, deptno,
    -> first_value(sal) over w first,
    -> last_value(sal) over w last,
    -> nth_value(sal, 3) over(partition by deptno order by sal) 3th_sal
    -> from emp
    -> window w as
    -> (
```

```
    ->     partition by deptno order by sal
    ->     rows between unbounded preceding and unbounded following
    -> );
```

ename	sal	deptno	first	last	3th_sal
MILLER	1300.00	10	1300.00	5000.00	NULL
CLARK	2450.00	10	1300.00	5000.00	NULL
KING	5000.00	10	1300.00	5000.00	5000.00
SMITH	800.00	20	800.00	3000.00	NULL
JONES	2975.00	20	800.00	3000.00	NULL
FORD	3000.00	20	800.00	3000.00	3000.00
JAMES	950.00	30	950.00	2850.00	NULL
WARD	1250.00	30	950.00	2850.00	1250.00
MARTIN	1250.00	30	950.00	2850.00	1250.00
TURNER	1500.00	30	950.00	2850.00	1250.00
ALLEN	1600.00	30	950.00	2850.00	1250.00
BLAKE	2850.00	30	950.00	2850.00	1250.00

12 rows in set (0.01 sec)

下面的示例以整个 emp 表作为一个窗口,查询每个行之前 2 行和之后 2 行的 sal 值。

```
mysql> select deptno, ename, sal,
    -> lead(sal, 2) over w,
    -> lag(sal, 2) over w
    -> from emp
    -> window w as (order by sal desc);
```

deptno	ename	sal	lead(sal, 2) over w	lag(sal, 2) over w
10	KING	5000.00	2975.00	NULL
20	FORD	3000.00	2850.00	NULL
20	JONES	2975.00	2450.00	5000.00
30	BLAKE	2850.00	1600.00	3000.00
10	CLARK	2450.00	1500.00	2975.00
30	ALLEN	1600.00	1300.00	2850.00
30	TURNER	1500.00	1250.00	2450.00
10	MILLER	1300.00	1250.00	1600.00
30	WARD	1250.00	950.00	1500.00
30	MARTIN	1250.00	800.00	1300.00
30	JAMES	950.00	NULL	1250.00
20	SMITH	800.00	NULL	1250.00

12 rows in set (0.00 sec)

第 8 章　数据修改语句

在诸多数据库操作中,数据修改功能的使用频率仅次于查询,也属于最常用的 SQL 语句。delete、update 及 insert 作为数据修改的 3 种语句,分别用于删除、更新及添加数据。执行这 3 种语句时,会同时产生重做数据及撤销数据。另外,MySQL 也支持兼具 insert 和 update 功能的 replace 语句。

本章主要内容包括:
- delete 语句;
- update 语句;
- insert 语句;
- replace 语句。

8.1　delete 语句

delete 语句用于删除表中的记录,一般会同时附带 where 子句限定删除记录的范围,如果不附加 where 子句,则会把整个表的数据清空。

8.1.1　简单的 delete 语句

删除 emp 表中的所有记录,即清空 emp 表中的数据:

```
mysql> delete from emp;
```

对 delete 语句附加 where 子句的方法与 select 语句相似,如删除 10 号部门的记录:

```
mysql> delete from emp where deptno = 10;
```

8.1.2　delete 语句使用 limit 子句

可以结合 order by 及 limit 子句,把行排序后,删除指定的前几行。
下面的命令为删除工资最高的前 2 行。

```
mysql> delete from emp order by sal desc limit 2;
```

与 select 语句中使用的 limit 子句不同,delete 语句的 limit 子句只支持一个参数。

8.1.3　delete 语句同时删除多个表的行

可以使用表连接方式同时删除多个表的行。
若多个表之间存在外键约束关联,则不适宜采用这种方式,因为这种方式可能会先删除主表记录,从而与外键约束冲突,导致操作失败。这种情况下,应对外键约束附加 on delete cascade 子句,在对主表记录执行删除操作时,子表记录会自动级联删除。或先删去外键,再使用这里的删除方法。

下面的示例用一个 delete 命令同时删去 emp 表和 dept 表中的记录。

先删去 emp 表上的外键。

```
mysql> alter table emp drop constraint fk_deptno;
```

下面的语句同时删除了 emp 表和 dept 表中满足多表连接条件和单表限制条件的行。

```
mysql> delete from e, d
    -> using emp e join dept d join salgrade s
    -> where e.deptno = d.deptno and e.sal between s.losal and s.hisal
    -> and s.grade = 1;
```

要注意,删除的行仅限于 from 子句中出现的表,using 子句中的表用于构造删除条件,另外,若使用表别名,需在 using 子句指定。

以上查询也可以不使用 using 子句,写为如下形式。

```
mysql> delete e, d
    -> from emp e join dept d join salgrade s
    -> where e.deptno = d.deptno and e.sal between s.losal and s.hisal
    -> and s.grade = 1;
```

连接条件中的 where 关键字也可以使用 on。

以上语法形式也支持外连接。

以下命令删除 dept 表中没有员工的行,即 deptno 列值未在 emp 表的 deptno 列出现。

```
mysql> delete d
    -> from emp e right outer join dept d
    -> on e.deptno = d.deptno
    -> where e.deptno is null;
```

8.2 update 语句

update 语句用于修改表中记录的列值,使用时一般会附加 where 子句,以限定被修改行的范围,否则会把整个表的指定列值全部修改。

8.2.1 简单的 update 语句

把 emp 表中 10 号部门的工资增加 10%。

```
mysql> update emp set sal = sal * 1.1 where deptno = 10;
```

可以用一个 update 语句同时修改多个列值,如把 20 号部门的 sal 列值改为 2 000,comm 列值改为 1 200。

```
mysql> update emp set sal = 2000, comm = 1200 where deptno = 20;
```

8.2.2 update 语句修改多个表

与 delete 操作多个表相似,一个 update 语句可以修改多个表的列值。

```
mysql> update emp e, dept d set e.sal = 1000, d.dname = 'DEV'
    -> where e.deptno = d.deptno and e.deptno = 10;
```

8.3 insert 语句

insert 语句用于向表添加记录，其主要语法包括 3 种不同的形式。
- insert into *table_name*(*column_list*) values(*value_list*)
- insert into *table_name* values(*value_list*)
- insert into *table1_name*(*column_list*) select *column_list* from *table2_name*

第 1 种形式：表名后附加列名列表，values 关键字后放置对应的列值，多个列值的顺序和类型要与前面列名列表一致。未在列名列表中出现的列，其值为 null。

第 2 种形式：表名后不放列名，而在 values 关键字后面输入整行记录的所有列值。若某个列为空，则要明确其值为 null，不能省略。

第 3 种形式：把一个表的数据导入另一个表。要求两个表的列个数和类型要一致。

下面通过几个实例说明每种语法形式的用法。

用第 1 种形式对 dept 表添加一行记录，第 3 个列未出现在列名列表，其值为 null。

```
mysql> insert into dept(deptno, dname) values(50,'HR');
```

用第 2 种形式对 dept 表添加一行记录，第 3 个列值为 null。

```
mysql> insert into dept values(60,'RD',null);
```

先创建一个测试表，然后对其用第 3 种语法形式添加记录：

```
mysql> create table t(a int, b char(10));
mysql> insert into t select deptno, dname from dept where deptno < 30;
```

8.4 replace 语句

若一个表没有主键约束也没有唯一索引，则 replace 语句的作用与 insert 语句相同。

若表上存在主键约束或唯一索引，而且新加行的主键或唯一索引列值与现有记录重复，则用新行替换旧行；若不重复，则与普通的 insert 语句相同。

```
mysql> select * from dept;
+--------+------------+----------+
| deptno | dname      | loc      |
+--------+------------+----------+
|     10 | ACCOUNTING | NEW YORK |
|     20 | RESEARCH   | DALLAS   |
|     30 | SALES      | CHICAGO  |
|     40 | OPERATIONS | BOSTON   |
+--------+------------+----------+
4 rows in set (0.00 sec)
mysql> replace dept values(40,'AI','PITTSBURGH'),(50,'BI','AUSTIN');
Query OK, 3 rows affected (0.00 sec)
Records: 2  Duplicates: 1  Warnings: 0

mysql> select * from dept;
```

```
+--------+------------+-----------+
| deptno | dname      | loc       |
+--------+------------+-----------+
|     10 | ACCOUNTING | NEW YORK  |
|     20 | RESEARCH   | DALLAS    |
|     30 | SALES      | CHICAGO   |
|     40 | AI         | PITTSBURGH|
|     50 | BI         | AUSTIN    |
+--------+------------+-----------+
5 rows in set (0.00 sec)
```

第 9 章 表 及 约 束

关系型数据库的数据存储在表里,表是最重要的数据库对象。在数据库设计完成后,得到表的逻辑结构,并根据实际情况,确定了列的类型及所需满足的约束条件,接下来就可以在数据库执行创建表的任务了。本章讲解创建表和管理表的 SQL 语句,也是 DDL 语句中比较常用的一类。

本章主要内容包括:
- 创建简单的表;
- 自增列和自动填充列;
- 字符集和排序规则;
- 约束;
- 复制表;
- 修改表的结构;
- 查询表的定义。

9.1 创建简单的表

简单的表指未附加约束的表,各种属性都取默认值,只需指定表名、列名及列类型。

创建表使用 create table 语句,下面命令创建表 t,包含两个列 a、b,其中 a 为整型,b 为定长字符串,最长 15 个字符。

```
mysql> create table t(a int, b char(15));
```

9.2 字符集及排序规则

字符集是某个范围内的所有字符的汇总,如中文字符、日文字符、英文字母等。计算机行业的字符集除了字符本身外,还包括每个字符对应的数字,以及数字以什么格式存储(如占用几个字节)。

ASCII 码是最早的字符集。随着计算机技术的发展,各个国家都设计了本国语言字符集,如中国的 GB 2312、GB 18030。一个国家的字符集很可能不包含另一个国家使用的字符,如中文字符集 GB 18130 不包含阿拉伯字符。当一个应用同时需要处理多国语言时,这些字符集的缺点就会马上暴露出来,解决方案是使用 Unicode 字符集。

字符集的另一个重要属性是排序规则,如英文字母是否区分大小写,中文是否按照拼音。

9.2.1 MySQL 与 Unicode 字符集

为了解决多种语言支持的问题,1991 年产生了第一个 Unicode 国际字符编码系统

(Unicode Standard 1.0),其目的是把世界上的所有字符包含进来。Unicode 编码分为 17 个部分或平面(官方名称为 Plane),每个部分包括 65 536 个字符。这些编码平面总体上只使用了 10% 左右,Plane 3 至 Plane 13 完全未使用。最重要的是第 1 部分,即 Plane 0,包括世界上几乎所有字符,称为 Basic Multilingual Plane,简写为 BMP。其他 16 个平面称为增补部分,用于存储很少使用的字符,如 emoji 表情符号(这些表情符号现在已经使用很频繁了,特别是☺)、罕见汉字等。

从提出以来,Unicode 标准一直在更新,字符数量一直在增加,2003 年的 Unicode 4.0 包含 96 382 个字符,2020 年的 Unicode 13.0 则增加至 143 859 个字符。

Unicode 编码系统规定了字符总量和字符与数字的对应关系,但一个数字如何存储(如用几个字节),有不同的实现方案,这些方案即各种不同的 Unicode 字符集。

查看 MySQL 支持的字符集可以执行 show character set,也可以使用 information_schema 数据库中的 character_sets 视图。

以下示例使用 show 命令查看 MySQL 支持的所有字符集。

```
mysql> show character set;
+----------+-----------------------+---------------------+--------+
| Charset  | Description           | Default collation   | Maxlen |
+----------+-----------------------+---------------------+--------+
| armscii8 | ARMSCII-8 Armenian    | armscii8_general_ci |      1 |
| ascii    | US ASCII              | ascii_general_ci    |      1 |
| big5     | Big5 Traditional Chinese | big5_chinese_ci  |      2 |
...
| utf8mb4  | UTF-8 Unicode         | utf8mb4_0900_ai_ci  |      4 |
+----------+-----------------------+---------------------+--------+
41 rows in set (0.00 sec)
```

附加条件,可以过滤出 MySQL 支持的所有 Unicode 字符集。

```
mysql> show character set where description like '%Unicode%';
+---------+----------------+--------------------+--------+
| Charset | Description    | Default collation  | Maxlen |
+---------+----------------+--------------------+--------+
| ucs2    | UCS-2 Unicode  | ucs2_general_ci    |      2 |
| utf16   | UTF-16 Unicode | utf16_general_ci   |      4 |
| utf16le | UTF-16LE Unicode | utf16le_general_ci |    4 |
| utf32   | UTF-32 Unicode | utf32_general_ci   |      4 |
| utf8    | UTF-8 Unicode  | utf8_general_ci    |      3 |
| utf8mb4 | UTF-8 Unicode  | utf8mb4_0900_ai_ci |      4 |
+---------+----------------+--------------------+--------+
6 rows in set (0.00 sec)
```

其中,ucs2 和 utf8 只支持 BMP 基本字符,其他几种都支持 BMP 基本字符和增补字符。

MySQL 8 的默认字符集为 utf8mb4,包括 Unicode 字符集中的 BMP 基本字符和增补字符,在字符包含范围方面一般不会有问题。

utf8mb4 中的 utf8 表示字符编码基本单位为一个字节,mb4 表示 max bytes 4,即一个字符最多分配 4 个字节存储。utf8mb4 字符集中的一个字符可以由 1~4 个字节存储,最常用的

ASCII 字符用 1 个字节，汉字一般用 3 个字节，个别生僻字符用 4 个字节。

utf16 的基本编码单位是 2 个字节，一个字符用 2 个或 4 个字节存储。utf32 的基本编码单位是 4 个字节，每个字符都用 4 个字节存储。

9.2.2 排序规则与中文排序

字符集的属性除了编码存储方式和字符范围外，另一个重要属性就是对字符集内的字符排序。每个 Unicode 字符集都支持多种文字，每种文字都有自己的特有排序方式，一个 Unicode 字符集也支持多种排序规则。

查看所有排序规则可以使用 show collation 命令或 information_schema.collations 视图。

使用 information_schema.collations 视图，查看 utf8mb4 支持的排序规则。

```
mysql> select collation_name from information_schema.collations
    -> where character_set_name = 'utf8mb4';
+----------------------------+
| collation_name             |
+----------------------------+
| utf8mb4_general_ci         |
| utf8mb4_bin                |
...
| utf8mb4_zh_0900_as_cs      |
| utf8mb4_0900_bin           |
+----------------------------+
75 rows in set (0.01 sec)
```

排序规则名称一般包含字符集名称，语言，排序算法版本号，是否区分重音，字母是否区分大小写。如 utf8mb4_zh_0900_as_cs 中，utf8mb4 为字符集名称；zh 为中文；0900 为排序算法版本号，即 9.0；as 表示区分重音（accent sensitive，即把 e、è、é、ê、ë 看作不同的字符）；cs 表示字母区分大小写（case sensitive，即把 A 和 a 看作不同的字符）。

utf8mb4 的默认排序规则为 utf8mb4_0900_ai_ci，不区分字母大小写，也不支持中文排序。如果需要区分字母大小写或支持中文拼音排序，可以选用 utf8mb4_zh_0900_as_cs，此排序规则为 MySQL 8.0.16 版本增加。

9.2.3 建表时设置字符集及排序规则

建表时，可以用 character set（简写为 charset）和 collate 子句指定字符集和排序规则。

```
mysql> create table t1(a int, b char(10))
    -> character set = utf8mb4
    -> collate = utf8mb4_zh_0900_as_cs;
```

若指定了排序规则，其名称中已包含字符集，character set 子句可以省略。

也可以对个别列指定排序规则。

```
mysql> create table t2
    ->(
    ->    a int,
    ->    b char(10) collate utf8mb4_zh_0900_as_cs
    ->);
```

除了以上设置方式,也可以设置服务器或某个数据库的默认字符集和排序规则。
在建库时,设置字符集和排序规则。

```
mysql> create database db
    -> character set = utf8mb4
    -> collate = utf8mb4_zh_0900_as_cs;
```

以下命令设置服务器的字符集和排序规则。

```
mysql> set persist character_set_server = utf8mb4;
mysql> set persist collation_server = utf8mb4_zh_0900_as_cs;
```

上面的设置需要用户重新登录后才会生效。

9.3 建表时指定存储引擎

MySQL 支持多种存储引擎,默认为 InnoDB,建表时可以指定其适用的存储引擎,因为 InnoDB 支持其他引擎不具备的事务处理、外键约束等特性,所以一般没有选择其他存储引擎的理由。除非必要,本书只介绍 InnoDB 引擎。

指定存储引擎为 InnoDB。

```
mysql> create table t3(a int, b char(10))
    -> engine = innodb;
```

9.4 使用 auto_increment 自增列

某些情况下,表的列都不足以唯一地区分一行,这时可以使用 auto_increment 选项对其添加一个自增列,如订单表的订单 id。对表添加记录时,不用理会自增列,其值默认从 1 开始,依次递增。这种自增列需要指定为主键或附加唯一约束。

下面示例中,t 表中的 a 列设置为自增列主键,然后对其添加记录。

```
mysql> create table t(a int auto_increment primary key, b char(10));
mysql> insert into t(b) values('hello'), ('world');
mysql> select * from t;
+---+-------+
| a | b     |
+---+-------+
| 1 | hello |
| 2 | world |
+---+-------+
2 rows in set (0.00 sec)
```

系统变量 auto_increment_offset 和 auto_increment_increment 用于设置自增列的初值和步长,默认均为 1。

查询自增列信息,可以使用系统视图 sys.schema_auto_increment_columns,其中的 auto_increment 列值为下一个应取值。

```
mysql> SELECT * FROM sys.schema_auto_increment_columns\G
*************************** 1. row ***************************
        table_schema: db
          table_name: t
         column_name: a
           data_type: int
         column_type: int
           is_signed: 1
         is_unsigned: 0
           max_value: 2147483647
      auto_increment: 4
auto_increment_ratio: 0.0000
1 row in set (0.00 sec)
```

9.5 自填充时间列

很多情况下，表会有一个时间列，用于存储行的生成时间或修改时间，如订单的生成时间、用户的注册时间、论坛帖子的发表时间等。这种时间数据不需要用户手工填充，一般只需要自动取到操作执行的时间即可。

在 MySQL 中，可以把这种列定义为 timestamp 或 datetime 类型，并将其 default 属性指定为当前时间，即可做到自填充。若再指定其 on update 属性为当前时间，则可做到自更新。

创建 users 表，其创建时间指定为自填充，修改时间指定为自填充和自更新。

```
mysql> create table users
    -> (
    ->     id int,
    ->     name char(10),
    ->     create_time timestamp default now(),
    ->     update_time timestamp default now() on update now()
    -> );
```

添加记录时，忽略自填充列。

```
mysql> insert into users(id, name) values(1,'Smith');
```

查询其内容，可以看到自填充列已有默认值。

```
mysql> select * from users;
+----+-------+---------------------+---------------------+
| id | name  | create_time         | update_time         |
+----+-------+---------------------+---------------------+
|  1 | Smith | 2021-02-08 16:33:26 | 2021-02-08 16:33:26 |
+----+-------+---------------------+---------------------+
1 row in set (0.00 sec)
```

若修改其行数据，则自更新时间列会自动更新其值。

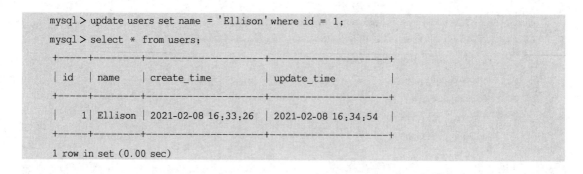

9.6 约 束

为了减少出现错误数据的可能性,可以在创建表时对其列附加约束,也可以在建表后,再增加约束。这里的主键约束和外键约束也是完成关系模型三要素中的第三部分,实现实体完整性约束和引用完整性约束。

9.6.1 约束的种类

与其他 DBMS 产品相似,MySQL 提供了下面几种约束。

- 主键约束(primary key):列值不能为空也不能重复。一个表的主键约束只能有一个。
- 唯一约束(unique):列上的值不能重复。
- 外键约束(foreign key):列值要匹配于主表的相应列值。外键所在的表称为子表。
- 检查约束(check):对列的取值范围附加限制条件。
- 非空约束(not null):列上的值不能为空。非空约束只能附加在列级。
- 默认约束(default):对表添加记录时,若未指定列值,则取其默认值。

MySQL 8.0.16 开始支持检查约束,之前版本在执行语法解析时不报错,但检查约束本身会略去。

在创建表时,根据语法形式,附加的约束可以分为以下两种类型。

- 列级约束:约束子句直接附加在列定义之后。
- 表级约束:所有列的定义结束后,再附加约束相关子句。

一般来说,一个约束既可以使用列级形式附加,也可以使用表级形式附加,两种约束对表所产生的限制作用是相同的。包含多个列的复合约束只能使用表级形式。

附加约束时,可以指定名称,也可以不指定。若未指定名称,则由 MySQL 自动生成。主键约束是一个例外,不管是否附加了名称,其名称总为 PRIMARY。

附加主键约束时,一般使用表级无名称形式。

附加检查、唯一、外键约束时,建议使用表级、指定名称的方式,以便于识别和管理。

非空、默认约束一般当作列属性,MySQL 不支持这两种约束的表级形式,也不能使用名称。

9.6.2 主键、非空及默认约束

不管用户附加约束时是否命名,MySQL 的主键约束名称总为 PRIMARY。MySQL 对主键约束自动创建聚集索引。

以下示例附加了表级主键约束及列级非空、默认约束。

```
mysql> create table p
    -> (
    ->     a int,
    ->     b int not null,
    ->     c int default 0,
    ->     primary key(a)
    -> );
```

为了与其他产品兼容,使用下面的表级指定名称的语法形式附加主键,MySQL 会在执行时解析通过,但名称其实不起作用。

```
mysql> create table p1
    -> (
    ->     a int,
    ->     b int,
    ->     constraint pk_p1 primary key(a)
    -> );
```

其他产品一般都支持的列级附加名称的主键语法形式,解析时不会通过。

```
mysql> create table p2
    -> (
    ->     a int constraint pk_p2 primary key,
    ->     b int
    -> );
ERROR 1064 (42000): You have an error in your SQL syntax; check the manual that corresponds to your MySQL server version for the right syntax to use near 'primary key,
    b int
)' at line 3
```

9.6.3 唯一、检查及外键约束

这几种约束一般使用表级附加名称的语法形式。

在下面的示例中,外键指向的 p 表为上一节创建。

```
mysql> create table c
    -> (
    ->     x int,
    ->     y int,
    ->     z int,
    ->     constraint uq_c unique(x),
    ->     constraint fk_c foreign key(y) references p(a),
    ->     constraint ck_c check(z > 0)
    -> );
```

查看以上建表语句在数据库中存储的对应定义。

```
mysql> show create table c \G
*************************** 1. row ***************************
       Table: c
Create Table: CREATE TABLE `c` (
  `x` int DEFAULT NULL,
  `y` int DEFAULT NULL,
  `z` int DEFAULT NULL,
  UNIQUE KEY `uq_c` (`x`),
  KEY `fk_c` (`y`),
  CONSTRAINT `fk_c` FOREIGN KEY (`y`) REFERENCES `p` (`a`),
  CONSTRAINT `ck_c` CHECK ((`z` > 0))
) ENGINE = InnoDB DEFAULT CHARSET = utf8mb4 COLLATE = utf8mb4_0900_ai_ci
1 row in set (0.00 sec)
```

在 MySQL 中，key 与 index 同义，唯一约束和外键约束都自动创建了索引。

在 c 表的 y 列上附加外键约束后，对子表 c 及主表 p 有如下限制：

- c 表中 y 列的值要匹配于 p 表中 a 列的值，不能超出其范围；
- p 表被 c 表的 y 列引用的记录不能删除，被引用的 a 列的值也不能修改。

在附加外键约束时，可以指定下面的子句，以规定当删除主表中被子表引用的记录或修改主表中被引用的列值时，如何自动处理子表中的对应记录。

- on delete cascade：删除主表中被引用的记录时，子表的对应记录级联删除。
- on update cascade：修改主表中被引用的列值时，子表的对应外键值级联修改。
- on delete / update set null：删除主表中被引用的记录，或修改主表中被引用的列值时，子表中的对应外键值设置为 null。
- on delete / update set default：删除主表中被引用的记录，或修改被引用的列值时，子表中的对应外键值设置为默认值。
- on delete / update no action，on delete / update restrict：no action 与 restrict 同义，其功能是禁止删除主表中被引用的记录，或禁止修改被引用的列值。如果未附加 on delete / update 子句，则此选项为默认行为。no action 的形式是为了兼容 SQL 标准，restrict 的形式是 MySQL 对 SQL 标准形式的扩展。如果使用 no action 形式，在存储表定义时，会略去，但 restrict 形式会保留。

下面是附加这几个子句的各种语法示例。

```
mysql> create table c1
    -> (
    ->     u int,
    ->     v int,
    ->     w int,
    ->     x int,
    ->     y int,
    ->     z int,
    ->     constraint fk_v foreign key(v) references p(a),
    ->     constraint fk_w foreign key(w) references p(a) on delete cascade,
    ->     constraint fk_x foreign key(x) references p(a) on update set null,
```

```
        ->       constraint fk_y foreign key(y) references p(a) on delete no action,
        ->       constraint fk_z foreign key(z) references p(a) on update restrict
        -> );
```

查看其存储的表定义。

```
mysql> show create table c1 \G
*************************** 1. row ***************************
       Table: c1
Create Table: CREATE TABLE `c1` (
  `u` int DEFAULT NULL,
  `v` int DEFAULT NULL,
  `w` int DEFAULT NULL,
  `x` int DEFAULT NULL,
  `y` int DEFAULT NULL,
  `z` int DEFAULT NULL,
  KEY `fk_v` (`v`),
  KEY `fk_w` (`w`),
  KEY `fk_x` (`x`),
  KEY `fk_y` (`y`),
  KEY `fk_z` (`z`),
  CONSTRAINT `fk_v` FOREIGN KEY (`v`) REFERENCES `p` (`a`),
  CONSTRAINT `fk_w` FOREIGN KEY (`w`) REFERENCES `p` (`a`) ON DELETE CASCADE,
  CONSTRAINT `fk_x` FOREIGN KEY (`x`) REFERENCES `p` (`a`) ON UPDATE SET NULL,
  CONSTRAINT `fk_y` FOREIGN KEY (`y`) REFERENCES `p` (`a`),
  CONSTRAINT `fk_z` FOREIGN KEY (`z`) REFERENCES `p` (`a`) ON UPDATE RESTRICT
) ENGINE = InnoDB DEFAULT CHARSET = utf8mb4 COLLATE = utf8mb4_0900_ai_ci
1 row in set (0.00 sec)
```

可以通过开启或关闭系统变量 foreign_key_checks 来设置是否启用外键约束,默认为 1,即开启,0 则关闭。开启此变量时,只检查新添加的数据,不检查现有表的数据。

9.6.4 对表增加约束

建表后添加约束与创建表时附加表级约束的语法相似。

alter table *table_name* add constraint *constraint_name constraint_clause*

主键约束不需指定名称,如对 salgrade 表的 grade 列增加主键约束:

```
mysql> alter table salgrade add primary key(grade);
```

对于唯一、检查以及外键这几种约束来说,增加约束的语法相似。如对 emp 表的 sal 列增加表级检查约束。

```
mysql> alter table emp add constraint ck_sal check(sal between 0 and 8000);
```

对于非空和默认约束,要通过修改列属性添加。如对 emp 表的 sal 列添加 not null 和 default 约束。

```
mysql> alter table emp modify sal numeric(7, 2) not null default 0;
```

9.6.5 删除约束

在 MySQL 8.0.19 之前的版本中，删除约束没有统一的语法形式。

如删除 t 表的外键约束。

```
mysql> alter table c1 drop foreign key fk_x;
```

删除主键约束。

```
mysql> alter table emp drop primary key;
```

删除检查约束。

```
mysql> alter table c drop check ck_c;
```

删除唯一约束，需要通过删除其索引实现。

```
mysql> drop index uq_c on c;
```

从 MySQL 8.0.19 开始，删除约束可以使用下面更通用且符合 SQL 标准的语法形式。

alter table *table_name* drop constraint *constraint_name*

删除 emp 表上的外键约束 fk_deptno。

```
mysql> alter table emp drop constraint fk_deptno;
```

主键约束的名称总为 PRIMARY，因 PRIMARY 为关键字，删除时要用反单引号标记（位于键盘的数字 1 键左侧）。

```
mysql> alter table emp drop constraint `primary`;
```

对于 not null 约束，可以通过修改列的类型为 null 将其删除，对于 default 约束，修改列属性时，不附加 default 选项即表示无默认值（或默认值为 null），下面的示例为删除 emp 表的 sal 列上的 not null 及 default 约束。

```
mysql> alter table emp modify sal numeric(7, 2) null;
```

9.6.6 查询约束的信息

show create table 命令可以查看表的定义，其中包括所有的约束信息。

```
mysql> show create table dept \G
*************************** 1. row ***************************
       Table: dept
Create Table: CREATE TABLE `dept` (
  `deptno` decimal(2,0) NOT NULL,
  `dname` varchar(20) DEFAULT NULL,
  `location` varchar(13) DEFAULT NULL,
  PRIMARY KEY (`deptno`)
) ENGINE = InnoDB DEFAULT CHARSET = utf8mb4 COLLATE = utf8mb4_0900_ai_ci
1 row in set (0.00 sec)
```

除此之外，系统数据库 information_schema 中的几个字典视图专门用来查询约束信息，这几个视图除了都包括约束所在表名称及数据库名称外，各有侧重点。

table_constraints：包括所有约束，但只提供名称和类型信息。其他几个视图不包括类型。

check_constraints：包括检查约束信息，可以查询其名称及检查条件。

referential_constraints：包括外键约束信息，可以查询其名称、指向的唯一约束名称及主表名称，以及对主表的 delete、update 规则，但未提供指向的列名。

key_column_usage:包括主键、唯一及外键约束信息。对于外键提供了其指向的列名。

如果查询某类约束,可以选择以上某个视图,如查询主键及外键约束信息。

```
mysql> select tc.table_name, tc.constraint_name cons_name,
    ->        tc.constraint_type cons_type, k.column_name col_name,
    ->        k.referenced_table_name || '(' || k.referenced_column_name || ')'
    ->        as reference
    -> from information_schema.table_constraints tc,
    ->      information_schema.key_column_usage k
    -> where tc.table_name = k.table_name and
    ->       tc.constraint_name = k.constraint_name and
    ->       tc.table_schema = 'law' and tc.table_name = 'emp';
+------------+-----------+-------------+----------+--------------+
| TABLE_NAME | cons_name | cons_type   | col_name | reference    |
+------------+-----------+-------------+----------+--------------+
| emp        | PRIMARY   | PRIMARY KEY | empno    | NULL         |
| emp        | fk_deptno | FOREIGN KEY | deptno   | dept(deptno) |
+------------+-----------+-------------+----------+--------------+
2 rows in set (0.00 sec)
```

9.7 复 制 表

MySQL 提供了两个语句用于复制表。

create table as select 语句通过查询命令得到新表的结构,包括查询语句中涉及的列及其类型,并把查询语句得到的记录加入新表,但除了非空和默认约束外,不会复制其他约束。

create table like 语句可以复制整个源表的结构,包括源表的所有列及其类型,除了外键约束外,其他约束也会复制过来,但不会复制源表的记录。

下面分别使用两个命令创建 t1 表和 t2 表。

```
mysql> create table t1 as select * from emp where 1 = 0;
mysql> create table t2 like emp;
```

第一个命令中的条件:where 1 = 0,目的是使查询结果为空,即只复制表结构,不复制其数据,去除此条件,则连同数据一起复制。

9.8 修改表的结构

修改表的结构主要包括修改列的数据类型,添加或删除列,添加或删除约束,修改表名等。完成这些任务要使用 alter table 语句。

9.8.1 修改列的数据类型

MySQL 使用 modify 关键字修改列的数据类型,其语法如下:

alter table *table_name* modify *column_name datatype*

如修改 dept 表的 dname 列为 varchar(20),可以执行以下命令:

```
mysql> alter table dept modify dname varchar(20);
```

若表尚不包含记录或列值都为 null,则可以对此列的类型、精度或长度做任意修改。

若列中已有非 null 值,则要注意:

- 修改列的精度或长度时,要求修改后的精度或长度能容纳现有数据;
- 修改列的类型时,要求新类型与原类型兼容,如 varchar 类型可以修改为 char 类型,但不能修改为 number 类型。

9.8.2 添加或删除列

添加或删除列分别使用以下语法。

- 添加列:alter table *table_name* add *column_name datatype*
- 删除列:alter table *table_name* drop column *column_name*

如对 dept 表添加列 phone_number,数据类型为 char(11)。

```
mysql> alter table dept add phone_number char(11);
```

删除以上的 phone_number。

```
mysql> alter table dept drop column phone_number;
```

9.8.3 修改列名

MySQL 使用 alter table 的 rename column 子句修改列名。

修改 dept 表的 loc 列为 location。

```
mysql> alter table dept rename column loc to location;
```

9.8.4 修改表名

MySQL 可以使用两种命令修改表的名称:

- alter table *table_name* rename to *new_name*
- rename table *table_name* to *new_name*

下面示例使用 alter table 命令修改 dept 表名称为 department,然后使用 rename table 命令将其改回原名。

```
mysql> alter table dept rename to department;
mysql> rename table department to dept;
```

9.8.5 清空表:truncate table

若执行 delete 语句时不附加 where 条件,则会清空表的数据。但是如果真的要清空表的数据,应该使用 truncate table 语句,而不是 delete 语句。

执行 truncate table 清空 emp 表。

```
mysql> truncate table emp;
```

与 delete 语句相比,truncate table 语句有以下特点。

- truncate table 语句属于 DDL 语句,执行后自动提交,不能回滚其效果,delete 属于 DML 语句,其删除的数据可以通过执行 rollback 命令恢复回来。
- truncate table 通过释放表占用的空间删除数据,虽然清空了表,但提示信息会显示影响的行数为 0,产生的重做数据很少。而通过 delete 删除的数据不会释放空间(再次添加记录时,会重用删除记录占用的空间),删除的行都会记入 undo 表空间和重做日志文件,产生的重做数据一般会比删除的数据量大,速度会较慢。

9.8.6 删除表

删除表使用 drop table 命令,通过以下命令删除 emp、dept 表。

```
mysql> drop table emp, dept;
```

9.9 查看表定义

如果只查看列的构成和列的类型,可以使用 desc 命令。

```
mysql> desc dept;
+--------+-------------+------+-----+---------+-------+
| Field  | Type        | Null | Key | Default | Extra |
+--------+-------------+------+-----+---------+-------+
| deptno | int         | NO   | PRI | NULL    |       |
| dname  | varchar(14) | YES  |     | NULL    |       |
| loc    | varchar(13) | YES  |     | NULL    |       |
+--------+-------------+------+-----+---------+-------+
3 rows in set (0.00 sec)
```

如果查看表的详细定义,可以使用 show create table 命令。

```
mysql> show create table emp\G
*************************** 1. row ***************************
       Table: emp
Create Table: CREATE TABLE `emp` (
  `empno` int NOT NULL,
  `ename` varchar(10) DEFAULT NULL,
  `job` varchar(9) DEFAULT NULL,
  `mgr` int DEFAULT NULL,
  `hiredate` datetime DEFAULT NULL,
  `sal` decimal(7,2) DEFAULT NULL,
  `comm` decimal(7,2) DEFAULT NULL,
  `deptno` int DEFAULT NULL,
  PRIMARY KEY (`empno`),
  KEY `fk_deptno` (`deptno`),
  CONSTRAINT `fk_deptno` FOREIGN KEY (`deptno`) REFERENCES `dept` (`deptno`)
) ENGINE = InnoDB DEFAULT CHARSET = utf8mb4 COLLATE = utf8mb4_0900_ai_ci
1 row in set (0.00 sec)
```

第 10 章　分　区　表

分区是数据文件中的特定区域,分区表以某个列为分区基准,添加的行根据其基准列值存入不同分区。以分区列作为查询条件时,根据条件限定的范围,只需访问特定的一个或几个分区,而不需要扫描整个表。分区表是优化查询的一种常用技术。

本章主要内容包括:
- 分区类别;
- 范围分区、列表分区、散列分区;
- 子分区;
- 验证执行计划使用了分区或子分区;
- 查询分区相关信息;
- 改变分区类型。

10.1　分区类别

根据划分方法的不同,分区可以分为以下几类。
- 范围分区(range):用不等式表示的范围作为分区依据。
- 列表分区(list):把分区列值常量作为分区依据。
- 散列分区(hash):对分区列执行散列后,以散列值作为分区依据。
- 键分区(key):散列分区的一种特殊情况,自动以主键列为分区列。

范围分区和列表分区支持单列及多列两种分区方法,单列分区只支持整数列作为分区列,多列分区除支持整型分区列外,还支持字符串及日期型列,但多列分区表不支持对分区列使用函数。散列分区只支持整型分区列。

主键约束和唯一约束都要求包含所有的分区列。

下面将说明各种分区的用法,为了使语法简洁,示例中未附加主键约束及唯一约束。

10.2　范围分区

范围分区以分区列值满足的范围作为行所属分区的依据。partition by range(*col*)子句表明分区类型为范围,values less than 子句把分区列值分为不同的连续范围。

10.2.1　单列范围分区

创建分区表 emps,分区列为 deptno。

```
mysql> create table emps
    -> (
    ->     empno numeric(4),
    ->     ename varchar(12),
    ->     hiredate datetime,
    ->     deptno int
    -> )
    -> partition by range(deptno)
    -> (
    ->     partition p0 values less than (20),
    ->     partition p1 values less than (40),
    ->     partition p2 values less than maxvalue
    -> );
```

maxvalue 表示无穷大。

以上命令对 emps 创建了 3 个分区 p0、p1 及 p2，其 deptno 的范围如下。

p0：deptno < 20

p1：21 ≤ deptno < 40

p2：41 ≤ deptno

如果 deptno 存在 null 值，MySQL 将其看作比任何非空值都小，相应行会落入 p0 分区。

查看执行计划，可以验证在不同的查询条件下只需扫描特定分区。如满足 deptno = 10 及 deptno = 50 的行分别在 p0 和 p2，相应查询只会扫描 p0 和 p2。

```
mysql> explain select * from emps where deptno = 10\G
*************************** 1. row ***************************
         id: 1
select_type: SIMPLE
      table: emps
 partitions: p0
       type: ALL
...
1 row in set, 1 warning (0.00 sec)


mysql> explain select * from emps where deptno = 50\G
*************************** 1. row ***************************
         id: 1
select_type: SIMPLE
      table: emps
 partitions: p2
       type: ALL
...
1 row in set, 1 warning (0.00 sec)
```

对分区列也可以使用返回值为整数的函数。

删除以上 emps 表后，对其重建，指定 year(hiredate) 为分区基准。

```
mysql> create table emps
    -> (
    ->     empno numeric(4),
    ->     ename varchar(12),
    ->     hiredate datetime,
    ->     deptno int
    -> )
    -> partition by range(year(hiredate))
    -> (
    ->     partition p0 values less than (2000),
    ->     partition p1 values less than (2010),
    ->     partition p2 values less than maxvalue
    -> );
```

以 hiredate = '2005-10-30' 为条件执行下面命令,通过执行计划可以验证相应查询只扫描了 p1 分区,也说明满足条件的行属于 p1 分区。

```
mysql> explain select * from emps where hiredate = '2005-10-30'\G
*************************** 1. row ***************************
           id: 1
  select_type: SIMPLE
        table: emps
   partitions: p1
         type: ALL
...
1 row in set, 1 warning (0.01 sec)
```

10.2.2 多列范围分区

多列范围分区使用 partition by range columns 子句指定多个分区列。

多列范围分区表的分区列也可以使用单列(多列的特例),可以避免单列范围分区的分区列必须为整型的限制。

对于多列范围分区表,使用行比较规则确定一个行的所属分区。两个行比较大小,先比较第 1 个列值,如果第 1 个列值不能确定,再比较第 2 个列值,依此类推。

如 (5, 10) < (6, 10),(5, 10) < (5, 11),(5, 10) = (5, 10)。

```
mysql> select (5, 11) < (6, 10),(5, 10) < (5, 11),(5, 10) = (5, 10);
+------------------+------------------+------------------+
| (5, 11) <(6, 10) | (5, 10) <(5, 11) | (5, 10) = (5, 10) |
+------------------+------------------+------------------+
|                1 |                1 |                1 |
+------------------+------------------+------------------+
1 row in set (0.00 sec)
```

下面示例以日期型 hiredate 列为分区列。

```
mysql> create table emps
    -> (
    ->      empno numeric(4),
    ->      ename varchar(12),
    ->      hiredate datetime,
    ->      deptno int
    -> )
    -> partition by range columns(hiredate)
    -> (
    ->      partition p0 values less than ('2000-01-01'),
    ->      partition p1 values less than ('2010-01-01'),
    ->      partition p2 values less than maxvalue
    -> );
```

下面示例使用 deptno 和 hiredate 作为分区列。

```
mysql> create table emps
    -> (
    ->      empno numeric(4),
    ->      ename varchar(12),
    ->      hiredate datetime,
    ->      deptno int
    -> )
    -> partition by range columns(deptno, hiredate)
    -> (
    ->      partition p0 values less than (20,'2000-01-01'),
    ->      partition p1 values less than (30,'2010-01-01'),
    ->      partition p2 values less than (maxvalue, maxvalue)
    -> );
```

确定一行所属的分区时，先以 deptno 为准，如果不能确定，再比较 hiredate 值，如 deptno = 10，hiredate = '2015-03-01'，只需比较 deptno 的范围即可确定此行属于 p0。

```
mysql> explain select * from emps
    -> where deptno = 10 and hiredate = '2015-03-01'
    -> \G
*************************** 1. row ***************************
           id: 1
  select_type: SIMPLE
        table: emps
   partitions: p0
         type: ALL
...
1 row in set, 1 warning (0.00 sec)
```

而对于 deptno = 20，hiredate = '2015-03-01'，只比较 deptno 不能确定其所属分区，还要继续比较 hiredate 才能确定其所属分区为 p1。

```
mysql> explain select * from emps
    -> where deptno = 20 and hiredate = '2015-03-01'
    -> \G
*************************** 1. row ***************************
           id: 1
  select_type: SIMPLE
        table: emps
   partitions: p1
         type: ALL
...
1 row in set, 1 warning (0.00 sec)
```

若查询条件只包含第 1 个分区列,且恰好为临界值,一般需要扫描多个分区,如 deptno ＝ 20,需要扫描 p0 和 p1 两个分区。

```
mysql> explain select * from emps
    ->  where deptno = 20
    -> \G
*************************** 1. row ***************************
           id: 1
  select_type: SIMPLE
        table: emps
   partitions: p0,p1
         type: ALL
...
1 row in set, 1 warning (0.00 sec)
```

若查询条件不包含第 1 个分区列,则优化器不能确定需要扫描的分区,那么需要扫描所有分区,即执行全表扫描。

```
mysql> explain select * from emps
    -> where hiredate = '2005-10-20'
    -> \G
*************************** 1. row ***************************
           id: 1
  select_type: SIMPLE
        table: emps
   partitions: p0,p1,p2
         type: ALL
...
1 row in set, 1 warning (0.00 sec)
```

由以上示例可以得出结论:对于多列范围分区表,如果查询时不使用第 1 个分区列作为条件,就失去分区的意义了。

10.2.3 增删范围分区

可以根据实际需要,增删分区。

alter table 命令使用 add partition 子句增加分区,使用 drop partition 子句删除分区。分

区删除后,其中的记录也被删除。若要删除分区,但保留记录,可以使用重组(合并)分区。

增加范围分区时,新的分区基准值要大于所有现有值,如果最后一个分区的分区基准为 less than maxvalue,则不能添加新分区了。如果新分区的基准值位于现有基准值之间,则可以通过重组(分割)现有分区实现。

创建 t 表作为测试表。

```
mysql> create table t(a int, b char(10))
    -> partition by range(a)
    -> (
    ->     partition p0 values less than(10),
    ->     partition p1 values less than(20)
    -> );
```

执行下面的命令增加 p2、p3 两个分区,新分区不能包含 p0、p1 中的值。

```
mysql> alter table t add partition
    -> (
    ->     partition p2 values less than(30),
    ->     partition p3 values less than(40)
    -> );
```

执行下面的命令删除 p1、p2 两个分区。

```
mysql> alter table t drop partition p1, p2;
```

10.2.4 重组分区

在保留全部数据的情况下,重组分区可以把一个分区划分为多个,也可以把多个相邻分区合并为 1 个。

执行下面的命令重建 t 表。

```
mysql> create table t(a int, b char(10))
    -> partition by range(a)
    -> (
    ->     partition p0 values less than(10),
    ->     partition p1 values less than(20),
    ->     partition p2 values less than maxvalue
    -> );
```

继续执行下面的命令分割 p2 分区为 p2、p3、p4。

```
mysql> alter table t reorganize partition p2 into
    -> (
    ->     partition p2 values less than(30),
    ->     partition p3 values less than(40),
    ->     partition p4 values less than maxvalue
    -> );
```

如果要删除分区,但需要保留其中的记录,则可以把相邻分区合并。如把上述 p2、p3、p4 重新合并为 p2。

```
mysql> alter table t reorganize partition p2,p3,p4 into
    -> (
    ->     partition p2 values less than maxvalue
    -> );
```

10.3 列表分区

列表分区使用 partition by list(*col*) 子句指定分区列，使用 values in() 子句指定分区基准常量值。列表分区不支持与范围分区类似的 maxvalue 用法，即不能指定一个分区把不属于其他分区的行都包括进去。添加行时，若列值不属于分区基准列值，则会报错。若分区列存在 null 值，则 null 值需要存在于列表中，否则也会报错。

10.3.1 单列列表分区

单列列表分区以整型常量值作为分区基准。

下面示例以 deptno 为分区列，其值为 10 或 20 的行存入 p1 分区，其值为 30 或 40 的行存入 p2 分区。

```
mysql> create table emps
    -> (
    ->     empno numeric(4),
    ->     ename varchar(12),
    ->     hiredate datetime,
    ->     deptno int
    -> )
    -> partition by list(deptno)
    -> (
    ->     partition p1 values in(10, 20),
    ->     partition p2 values in(30, 40)
    -> );
```

deptno 列值不属于以上 4 个分区基准值的行，不能添入表中。

通过执行计划，可以验证，若条件为 deptno = 30，则此查询只扫描 p2 分区。

```
mysql> explain select * from emps where deptno = 30\G
*************************** 1. row ***************************
           id: 1
  select_type: SIMPLE
        table: emps
   partitions: p2
         type: ALL
...
1 row in set, 1 warning (0.00 sec)
```

10.3.2 多列列表分区

多列列表分区使用 list columns 子句指定分区列。下面示例以 deptno、job 为分区列，用不同的常量值组合分为 3 个分区。

```
mysql> create table emps
    -> (
    ->     empno numeric(4),
    ->     ename varchar(12),
    ->     job varchar(20),
    ->     deptno int
    -> )
    -> partition by list columns(deptno, job)
    -> (
    ->     partition p1 values in ((10,'SALES'),(20,'DEV')),
    ->     partition p2 values in ((20,'ADMIN'),(30,'RESEARCH')),
    ->     partition p3 values in ((40,'RESEARCH'))
    -> );
```

若条件恰好满足分区基准值,则查询只扫描一个分区,如条件 where deptno = 20 and job = 'ADMIN'。

```
mysql> explain select * from emps where deptno = 20 and job = 'ADMIN'\G
*************************** 1. row ***************************
           id: 1
  select_type: SIMPLE
        table: emps
   partitions: p2
         type: ALL
...
1 row in set, 1 warning (0.00 sec)
```

若条件只包含第 1 个分区列,则优化器会通过比较基准值确定需要扫描的分区,如 where deptno = 10 会扫描 p1 分区, where deptno = 20 会扫描 p1 和 p2 分区。

```
mysql> explain select * from emps where deptno = 10\G
*************************** 1. row ***************************
           id: 1
  select_type: SIMPLE
        table: emps
   partitions: p1
         type: ALL
...
1 row in set, 1 warning (0.00 sec)

mysql> explain select * from emps where deptno = 20\G
*************************** 1. row ***************************
           id: 1
  select_type: SIMPLE
        table: emps
   partitions: p1,p2
```

```
        type: ALL
...
1 row in set, 1 warning (0.00 sec)
```

若条件不包含第 1 个分区列,则会扫描所有分区,即执行全表扫描。

```
mysql> explain select * from emps where job = 'ADMIN'\G
*************************** 1. row ***************************
           id: 1
  select_type: SIMPLE
        table: emps
   partitions: p1,p2,p3
         type: ALL
...
1 row in set, 1 warning (0.00 sec)
```

或者条件虽包含第 1 个分区列,但不包含分区基准值,也会执行全表扫描。

```
mysql> explain select * from emps where deptno = 20 and job = 'CLERK'\G
*************************** 1. row ***************************
           id: 1
  select_type: SIMPLE
        table: emps
   partitions: NULL
         type: ALL
...
1 row in set, 1 warning (0.00 sec)
```

因为 deptno = 20 and job = 'CLERK' 不满足任何分区基准,所以这种行不能被添入表中。

多列列表分区也可以使用一个分区列,这可以避开单列列表分区表只能使用整型列分区的限制。下面的示例使用 job 列为分区列。

```
mysql> create table emps
    -> (
    ->     empno numeric(4),
    ->     ename varchar(12),
    ->     job varchar(20),
    ->     deptno int
    -> )
    -> partition by list columns(job)
    -> (
    ->     partition p1 values in ('SALES','DEV'),
    ->     partition p2 values in ('ADMIN','RESEARCH'),
    ->     partition p3 values in ('CLERK')
    -> );
```

10.3.3 增删列表分区

与增删范围分区相同,alter table 命令使用 add partition 子句增加列表分区,使用 drop

partition 子句删除列表分区。分区删除后,其中的记录也被删除。

创建 t 表作为测试表。

```
mysql> create table t(a int, b char(10))
    -> partition by list(a)
    -> (
    ->     partition p0 values in(5, 10),
    ->     partition p1 values in(15, 20)
    -> );
```

执行下面的命令增加 p2、p3 两个分区,新分区不能包含 p0、p1 中的值。

```
mysql> alter table t add partition
    -> (
    ->     partition p2 values in(25, 30),
    ->     partition p3 values in(35, 40)
    -> );
```

执行下面的命令删除 p1、p2 两个分区。

```
mysql> alter table t drop partition p1, p2;
```

10.3.4 重组列表分区

重组列表分区即把若干个分区的列表值重新划分,重组后的分区个数与重组前的个数无关,如把 p1、p2 重组至 p1、p2、p3,是把 p1 和 p2 中的列表基准值重新划分至 p1、p2、p3。

执行下面的命令重建 t 表。

```
mysql> create table t(a int, b char(10))
    -> partition by list(a)
    -> (
    ->     partition p0 values in(5, 10),
    ->     partition p1 values in(15, 20),
    ->     partition p2 values in(25, 30)
    -> )
    -> ;
```

继续执行下面的命令把 p0 和 p1 重组。

```
mysql> alter table t reorganize partition p0, p1 into
    -> (
    ->     partition p0 values in(10, 15),
    ->     partition p1 values in(5),
    ->     partition p3 values in(20)
    -> );
```

10.4 散列分区

创建散列分区表时,只需指定分区列和分区个数。可以对分区列使用函数。MySQL 对每行的分区列值以散列算法决定其所属分区。如果表中的列没有明显特征,不适合使用范围分区或列表分区,则可以考虑使用散列分区。散列分区把 null 值看作 0。

10.4.1 普通散列分区

散列分区表用 partition by hash(*col*) 子句指定分区列,以 partitions 关键字指定分区个数。

```
mysql> create table emps
    -> (
    ->     empno numeric(4),
    ->     ename varchar(12),
    ->     job varchar(20),
    ->     deptno int
    -> )
    -> partition by hash(deptno) partitions 4;
```

散列算法使用简单的取余方式:以分区列值除以分区个数,取其余数,把余数相同的存储至同一个分区,余数值即为分区编号。如 mod(10,4) = 2,deptno = 10 的行会存入 p2 分区,而 mod(12,4) = 0,deptno = 12 的行会存入 p0 分区,如下面的执行计划所示。

```
mysql> explain select * from emps where deptno = 10\G
*************************** 1. row ***************************
           id: 1
  select_type: SIMPLE
        table: emps
   partitions: p2
         type: ALL
...
1 row in set, 1 warning (0.00 sec)

mysql> explain select * from emps where deptno = 12\G
*************************** 1. row ***************************
           id: 1
  select_type: SIMPLE
        table: emps
   partitions: p0
         type: ALL
...
1 row in set, 1 warning (0.00 sec)
```

10.4.2 线性散列分区

除了用取余算法进行散列之外,MySQL 还支持线性散列算法。其优点是在添加和删除分区以及合并和分割分区时,效率更高。缺点是其数据在各分区的分布不均衡。

线性散列分区表使用 partition by linear hash(*col*) 子句指定分区列。下面示例以 deptno 为分区列,使用线性散列算法创建分区表。

```
mysql> create table emps
    -> (
    ->     empno numeric(4),
    ->     ename varchar(12),
    ->     job varchar(20),
    ->     deptno int
    -> )
    -> partition by linear hash(deptno) partitions 4;
```

确定一行记录所属分区时,使用下面的算法:

$v = \text{power}(2, \text{ceiling}(\log(2, num)))$

$n = f(col)\ \&\ (v-1)$

while $n >= num$:

    set $v = v / 2$

    set $n = n\ \&\ (v-1)$

其中:num 为分区个数,n 为分区序号,$f(col)$ 为分区列值(或被函数作用的结果)。

如确定以上 emps 表中 deptno = 32 的行所属的分区过程如下:

$v = \text{power}(2, \text{ceiling}(\log(2, 32))) = \text{power}(2, \text{ceiling}(5)) = \text{power}(2, 5) = 32$

$n = 32\ \&\ (32-1) = 32\ \&\ 31 = 0$

最后得到其所属分区为 p0,可以由下面的执行计划验证。

```
mysql> explain select * from emps where deptno = 32\G
*************************** 1. row ***************************
           id: 1
  select_type: SIMPLE
        table: emps
   partitions: p0
         type: ALL
...
1 row in set, 1 warning (0.00 sec)
```

10.4.3 键分区

键(key)分区是散列分区的一种特殊情况,使用 partition by key()或 partition by linear key()子句。其分区列为主键列或附带 not null 的唯一约束列,不需要手工指定;如无主键或非空唯一约束,则需手工指定。

指定 empno 为主键,并创建键分区表。

```
mysql> create table emps
    -> (
    ->     empno numeric(4) primary key,
    ->     ename varchar(12),
    ->     job varchar(20),
    ->     deptno int
    -> )
    -> partition by key() partitions 4;
```

10.4.4 重组散列分区

散列分区不能像范围分区和列表分区一样增加或删除指定分区,但可以重组分区,即指定增加或减少分区个数。重组散列分区后,表中的记录会被重新存入合适的分区,但不会从表中删除。散列分区、键分区以及线性散列分区、线性键分区的重组语法相同,下面以普通散列分区为例说明其用法。

执行下面的命令创建散列分区表 t,指定其分区个数为 4。

```
mysql> create table t(a int, b char(10))
    -> partition by hash(a) partitions 4;
```

将其分区个数增加 8 个。

```
mysql> alter table t add partition partitions 8;
```

将其分区个数减少 4 个。

```
mysql> alter table t coalesce partition 4;
```

10.5 子 分 区

子分区是在范围分区或列表分区内再划分的分区,也称复合分区。子分区只能是散列分区或键分区(包括线性散列分区和线性键分区)。

子分区使用 subpartition by 子句定义相关属性,用法与 partition by 相似,可以根据需要指定子分区名称,若不指定名称,则由 MySQL 自动命名。在范围分区和列表分区内划分子分区的语法相似,本节以范围分区内划分散列子分区为例说明。

在范围分区中创建散列分区,只指定子分区个数。

```
mysql> create table emps
    -> (
    ->      empno numeric(4),
    ->      ename varchar(12),
    ->      hiredate datetime,
    ->      deptno int
    -> )
    -> partition by range(year(hiredate))
    -> subpartition by hash(deptno) subpartitions 2
    -> (
    ->      partition p0 values less than (2000),
    ->      partition p1 values less than (2010),
    ->      partition p2 values less than maxvalue
    -> );
```

查看执行计划,可以看到 MySQL 是如何命名,如何使用子分区的。

```
mysql> explain select * from emps where hiredate = '1999-10-25'\G
*************************** 1. row ***************************
           id: 1
  select_type: SIMPLE
        table: emps
```

```
       partitions: p0_p0sp0,p0_p0sp1
           type: ALL
...
1 row in set, 1 warning (0.01 sec)
```

若指定子分区的名称,可以使用下面的命令。

```
mysql> create table emps
    -> (
    ->      empno numeric(4),
    ->      ename varchar(12),
    ->      hiredate datetime,
    ->      deptno int
    -> )
    -> partition by range(year(hiredate))
    -> subpartition by hash(deptno) subpartitions 2
    -> (
    ->      partition p0 values less than (2000)
    ->      (
    ->          subpartition s00,
    ->          subpartition s01
    ->      ),
    ->      partition p1 values less than (2010)
    ->      (
    ->          subpartition s10,
    ->          subpartition s11
    ->      ),
    ->      partition p2 values less than maxvalue
    ->      (
    ->          subpartition s20,
    ->          subpartition s21
    ->      )
    -> );
```

10.6 查询分区信息

查看建表语句可以查看表的分区信息。通过 information_schema.partitions 视图可以查到更详细的信息。

以上节创建的复合分区表 emps 为例,查询其建表语句,注意其分区信息。

```
mysql> show create table emps\G
*************************** 1. row ***************************
       Table: emps
Create Table: CREATE TABLE `emps` (
  `empno` decimal(4,0) DEFAULT NULL,
  `ename` varchar(12) DEFAULT NULL,
```

```
  `hiredate` datetime DEFAULT NULL,
  `deptno` int DEFAULT NULL
) ENGINE = InnoDB DEFAULT CHARSET = utf8mb4 COLLATE = utf8mb4_0900_ai_ci
/* !50100 PARTITION BY RANGE (year(`hiredate`))
SUBPARTITION BY HASH (`deptno`)
(PARTITION p0 VALUES LESS THAN (2000)
 (SUBPARTITION s00 ENGINE = InnoDB,
  SUBPARTITION s01 ENGINE = InnoDB),
 PARTITION p1 VALUES LESS THAN (2010)
 (SUBPARTITION s10 ENGINE = InnoDB,
  SUBPARTITION s11 ENGINE = InnoDB),
 PARTITION p2 VALUES LESS THAN MAXVALUE
 (SUBPARTITION s20 ENGINE = InnoDB,
  SUBPARTITION s21 ENGINE = InnoDB)) */
1 row in set (0.00 sec)
```

使用 information_schema.partitions 系统视图查询 emps 的分区信息。

```
mysql> select partition_name, partition_method, partition_expression,
    -> subpartition_name, subpartition_method, subpartition_expression
    -> from information_schema.partitions
    -> where table_schema = 'db' and table_name = 'emps'
    -> \G
*************************** 1. row ***************************
         PARTITION_NAME: p0
       PARTITION_METHOD: RANGE
   PARTITION_EXPRESSION: year(`hiredate`)
      SUBPARTITION_NAME: s00
    SUBPARTITION_METHOD: HASH
SUBPARTITION_EXPRESSION: `deptno`
*************************** 2. row ***************************
         PARTITION_NAME: p0
       PARTITION_METHOD: RANGE
   PARTITION_EXPRESSION: year(`hiredate`)
      SUBPARTITION_NAME: s01
    SUBPARTITION_METHOD: HASH
SUBPARTITION_EXPRESSION: `deptno`
*************************** 3. row ***************************
         PARTITION_NAME: p1
       PARTITION_METHOD: RANGE
   PARTITION_EXPRESSION: year(`hiredate`)
      SUBPARTITION_NAME: s10
    SUBPARTITION_METHOD: HASH
SUBPARTITION_EXPRESSION: `deptno`
*************************** 4. row ***************************
```

```
         PARTITION_NAME: p1
       PARTITION_METHOD: RANGE
   PARTITION_EXPRESSION: year(`hiredate`)
      SUBPARTITION_NAME: s11
    SUBPARTITION_METHOD: HASH
SUBPARTITION_EXPRESSION: `deptno`
*************************** 5. row ***************************
         PARTITION_NAME: p2
       PARTITION_METHOD: RANGE
   PARTITION_EXPRESSION: year(`hiredate`)
      SUBPARTITION_NAME: s20
    SUBPARTITION_METHOD: HASH
SUBPARTITION_EXPRESSION: `deptno`
*************************** 6. row ***************************
         PARTITION_NAME: p2
       PARTITION_METHOD: RANGE
   PARTITION_EXPRESSION: year(`hiredate`)
      SUBPARTITION_NAME: s21
    SUBPARTITION_METHOD: HASH
SUBPARTITION_EXPRESSION: `deptno`
6 rows in set (0.00 sec)
```

10.7 改变表的分区类型

表的分区类型可以根据需要改变,或不使用分区。修改分区类型的语法与创建时的语法相似。

创建范围分区表 t 作为测试表。

```
mysql> create table t(a int, b int, c char(10))
    -> partition by range(a)
    -> (
    ->     partition p0 values less than(10),
    ->     partition p1 values less than(20),
    ->     partition p2 values less than maxvalue
    -> );
```

将其修改为列表分区,分区列为 b。

```
mysql> alter table t partition by list(b)
    -> (
    ->     partition p0 values in(1, 2, 3),
    ->     partition p1 values in(4, 5, 6)
    -> );
```

再将其修改为散列分区,分区个数为 4。

```
mysql> alter table t partition by hash(a) partitions 4;
```

执行下面的命令可以去除分区功能。

```
mysql> alter table t remove partitioning;
```

10.8 在 SQL 命令中直接操作分区

操作分区表时,一般由优化器决定操作的分区,用户也可以手工指定。

创建范围分区测试表 t。

```
mysql> create table t(a int, b int, c char(10))
    -> partition by range(a)
    -> (
    ->     partition p0 values less than(10),
    ->     partition p1 values less than(20),
    ->     partition p2 values less than maxvalue
    -> );
```

下面几个命令对 p0 分区执行 select、update 及 truncate 操作。

```
mysql> select * from t partition(p0) where a = 1;
Empty set (0.01 sec)


mysql> update t partition(p0) set b = 10;
Query OK, 0 rows affected (0.00 sec)
Rows matched: 0  Changed: 0  Warnings: 0


mysql> alter table t truncate partition p0;
Query OK, 0 rows affected (0.04 sec)
```

第 11 章 程序设计

MySQL 5.0 开始支持存储程序,目的是给面向集合的 SQL 语言加入面向过程的功能,可以在操作数据库时,使用变量、条件转向以及循环等面向过程语言的特征编写存储过程、函数、触发器和调度事件等对象,这 4 类对象的实现代码存储在数据库中,统称为存储程序或可编程对象,本书只讨论多种数据库产品通用的存储过程、函数、触发器 3 种对象。

本章主要内容包括:
- 用户变量的概念;
- MySQL 程序语言基本语法;
- 编写存储过程、函数、触发器;
- 查看存储程序系统信息;
- 删除存储程序。

11.1 用户变量

用户变量由 MySQL 3 引入,用于在一个会话内的语句间传递数据。虽然存储程序可以使用用户变量,但这不是引入它的主要目的,存储程序主要使用普通变量。使用用户变量时不需定义类型,其类型随其赋值自动转换。用户变量的名称以@作为标记,使用 set 语句或 select 语句执行赋值操作。

下面的示例把用户变量@a 赋值为 10,然后将其作为查询条件。

```
mysql> set @a = 10;
Query OK, 0 rows affected (0.00 sec)

mysql> select dname from dept where deptno = @a;
+------------+
| dname      |
+------------+
| ACCOUNTING |
+------------+
1 row in set (0.00 sec)
```

下面的示例把 emp 表的最高工资赋值给用户变量@a,然后查询获得最高工资的员工名称。

```
mysql> select max(sal) into @a from emp;
Query OK, 1 row affected (0.01 sec)

mysql> select ename from emp where sal = @a;
```

```
+-------+
| ename |
+-------+
| KING  |
+-------+
1 row in set (0.00 sec)
```

11.2 存储过程

存储过程是命名的程序块,程序块即其定义主体。与表一样,存储过程属于数据库内的对象。可以对用户赋予存储过程的执行权限。创建后,其定义存储于数据库,执行时,只需指定名称和参数。

本节利用存储过程说明 MySQL 的编程语法。MySQL 不支持 print 语句输出结果至屏幕,若需要输出结果,则可以使用 select 语句替代。函数和触发器的语法形式与本节说明相似。

11.2.1 存储过程的创建和执行

下面的示例说明存储过程从创建到执行的基本步骤。

程序语句要求以";"结尾,而";"本是执行 SQL 命令的标记,为了避免冲突,在创建存储过程之前,需要先以 delimiter 命令改变执行标记,如在下面的示例中改为"/"。

创建一个简单的存储过程,功能是输出 MySQL 的版本号。

```
mysql> delimiter /
mysql> create procedure show_version()
    -> select version() as 'MySQL Version';
    -> /
Query OK, 0 rows affected (0.02 sec)
```

存储过程创建完毕后,再次执行 delimiter 命令,把执行标记改回";",最后以 call 命令调用存储过程。

```
mysql> delimiter ;
mysql> call show_version();
+---------------+
| MySQL Version |
+---------------+
| 8.0.23        |
+---------------+
1 row in set (0.00 sec)
```

11.2.2 使用变量

MySQL 用下面 3 种语句处理变量。

- declare 语句:声明变量类型。
- set 语句:对变量赋值。
- select 语句:输出变量的值。

若程序主体由多条语句构成,则要用 begin … end 将其括起来。
下面是一个使用变量的简单示例。

```
mysql> delimiter /
mysql> create procedure get_sum()
    -> begin
    ->     declare n1, n2, s int;
    ->     set n1 = 12;
    ->     set n2 = 23;
    ->     set s = n1 + n2;
    ->     select s;
    -> end
    -> /
Query OK, 0 rows affected (0.02 sec)

mysql> delimiter ;
mysql> call get_sum();
+-----+
| s   |
+-----+
| 35  |
+-----+
1 row in set (0.00 sec)
```

11.2.3 使用 if 语句

if 语句用于流程控制中的条件转向。下面的示例根据当前时间的范围输出 'morning' 'noon' 或 'afternoon or night'。

```
mysql> delimiter /
mysql> create procedure show_part_of_day()
    -> begin
    ->         declare cur_time time;
    ->         declare  day_part varchar(30);
    ->         set cur_time = now();
    ->         if cur_time <'12:00:00' then
    ->                 set day_part = 'morning';
    ->         elseif cur_time = '12:00:00' then
    ->                 set day_part = 'noon';
    ->         else
    ->                 set day_part = 'afternoon or night';
    ->         end if;
    ->         select cur_time, day_part;
    -> end
    -> /
Query OK, 0 rows affected (0.01 sec)
```

```
mysql> delimiter ;
mysql> call show_part_of_day();
+----------+----------+
| cur_time | day_part |
+----------+----------+
| 11:21:44 | morning  |
+----------+----------+
1 row in set (0.00 sec)
```

11.2.4 使用 while 循环语句

MySQL 的循环结构支持 loop、repeat 及 while 语句,这里以求前 100 个自然数之和为例说明 while 语句的用法。

```
mysql> delimiter /
mysql> create procedure get_100_sum()
    -> begin
    ->     declare i, s int;
    ->     set i = 1;
    ->     set s = 0;
    ->     while i <= 100 do
    ->         set s = s + i;
    ->         set i = i + 1;
    ->     end while;
    ->     select s as 'Sum is';
    -> end
    -> /
Query OK, 0 rows affected (0.01 sec)

mysql> delimiter ;
mysql> call get_100_sum();
+--------+
| Sum is |
+--------+
|   5050 |
+--------+
1 row in set (0.00 sec)
```

11.2.5 使用输入参数及 SQL 语句

在存储过程代码中可以直接嵌入 SQL 命令,与 MySQL 数据库进行数据交互。

下面的示例通过创建存储过程得到指定部门的总工资,使用输入参数指定部门编号。

```
mysql> delimiter /
mysql> create procedure get_sumsal_by_dno(dno int)
    -> begin
    ->     declare s numeric(8,2);
    ->     select sum(sal) into s from emp where deptno = dno;
    ->     select concat('Total sal of ', dno, ' is ', s);
    -> end
    -> /
Query OK, 0 rows affected (0.00 sec)
```

输入参数指定为 10，执行存储过程。

```
mysql> delimiter ;
mysql> call get_sumsal_by_dno(10);
+------------------------------------+
| concat('Total sal of ', dno, ' is ', s) |
+------------------------------------+
| Total sal of 10 is 8750.00         |
+------------------------------------+
1 row in set (0.00 sec)
```

11.2.6 使用输出参数

存储过程没有返回值，函数有返回值，可以用输出参数模拟函数返回值的效果。输出参数需要在参数名称前指定 out。

在下面示例创建的存储过程中，s 参数为输出参数，用来存储两个输入参数 n1 和 n2 的和。

```
mysql> delimiter /
mysql> create procedure test_out(n1 int, n2 int, out s int)
    -> begin
    ->     set s = n1 + n2;
    -> end
    -> /
Query OK, 0 rows affected (0.00 sec)
```

调用存储过程，指定两个输入参数为 12 和 23，用户变量@a 用于接收输出参数的值。

```
mysql> delimiter ;
mysql> call test_out(12, 23, @a);
Query OK, 0 rows affected (0.00 sec)
```

最后显示其结果。

```
mysql> select @a;
+------+
| @a   |
+------+
|   35 |
+------+
1 row in set (0.00 sec)
```

11.3 函　　数

函数与存储过程的主要区别在于函数有返回值。定义函数时,需要指定返回值类型。另外,函数需要在 SQL 语句中调用,不能像存储过程一样单独调用。

为了执行函数时,优化器能更好地制订执行计划,在创建函数时,需要指定下面属性之一。
- deterministic:若输入参数相同,返回结果也相同,则使用此选项。否则使用 not deterministic,如生成随机数或返回当前时间。若省略,则默认为 not deterministic。
- contains sql:代码中包含 SQL 语句,但不涉及读取和修改数据库数据,如 lock tables。
- no sql:代码中不包含 SQL 语句。
- reads sql data:代码中包含读取数据的 SQL 语句。
- modifies sql data:代码中包含修改数据的 SQL 语句。

下面示例创建的函数返回两个定点小数之和,如果传入参数为 null,则将其看作 0。

```
mysql> delimiter /
mysql> create function sum_of_sal_comm(n1 numeric(7,2), n2 numeric(7,2))
    -> returns numeric(8,2) no sql
    -> begin
    ->     declare s numeric(8,2);
    ->     set s = ifnull(n1, 0) + ifnull(n2, 0);
    ->     return s;
    -> end
    -> /
Query OK, 0 rows affected (0.01 sec)
```

在查询语句中使用以上函数返回部门编号为 30 的员工工资和佣金之和。

```
mysql> delimiter ;
mysql> select ename, sal, comm, sum_of_sal_comm(sal, comm)
    -> from emp where deptno = 30;
+--------+---------+---------+----------------------------+
| ename  | sal     | comm    | sum_of_sal_comm(sal, comm) |
+--------+---------+---------+----------------------------+
| ALLEN  | 1600.00 |  300.00 |                    1900.00 |
| WARD   | 1250.00 |  500.00 |                    1750.00 |
| MARTIN | 1250.00 | 1400.00 |                    2650.00 |
| BLAKE  | 2850.00 |    NULL |                    2850.00 |
| TURNER | 1500.00 |    0.00 |                    1500.00 |
| JAMES  |  950.00 |    NULL |                     950.00 |
+--------+---------+---------+----------------------------+
6 rows in set (0.00 sec)
```

MySQL 未提供将字符串首字母大写、其他字母小写的函数,下面的函数能够完成这一功能。

```
mysql> create function initcap(s varchar(255))
    -> returns varchar(255) deterministic
    -> return concat(upper(left(s, 1)), lower(mid(s, 2)));
Query OK, 0 rows affected (0.03 sec)

mysql> select initcap('SMITH');
+------------------+
| initcap('SMITH') |
+------------------+
| Smith            |
+------------------+
1 row in set (0.02 sec)
```

11.4 触发器

触发器是与表关联的存储过程，其功能是对表的 insert、update、delete 操作执行约束检查或审计(audit)。触发器可以引用修改之前和之后的行，比普通的表约束功能更强。审计即记录下某些敏感操作的属性(如修改工资或考核成绩时的用户名、执行时刻等)，当有人在执行非法操作时，可以有迹可循。

对表执行 insert、update、delete 操作时，相应触发器会自动激活执行。触发器可设置为在执行 insert、update、delete 之前或之后被激活。

11.4.1 创建触发器的语法

创建触发器的语法如下：
create trigger *tri_name*
{before | after}
{insert | update | delete}
on *table_name*
for each row
begin
    statements
end

{before | after}：二者选一，指定触发器执行的时机。before 设置在触发动作执行之前激活触发器，触发器中的代码执行后，继续执行触发动作。after 设置在触发动作完成后才激活触发器。

{insert | update | delete}：三者选一，指定触发操作。

on *table_name*：指定触发器所属的表。

for each row：对触发动作影响的每一行都执行一次触发器，当前只支持这一种选项(但不能省略)。未来版本可能支持的另一种选项是对每一个语句执行一次触发器，即 for each statement。

begin … end：触发器主体代码。

在触发器中不能执行查询。为防止递归执行,触发器内也不能对本表执行 insert、update、delete 操作。

11.4.2 old 和 new 的用法

在触发器中,old 和 new 用来表示 update 操作修改之前的原行和修改之后的新行。引用更新前后的列值,如 a 列,可以使用 old.a 和 new.a。另外,insert 操作添加的新行在触发器中用 new 表示,delete 操作删除的行在触发器中用 old 表示。

对 emp 表创建触发器,考查一下 old 和 new 的内容,以理解其含义。因为触发器中不能使用查询,所以我们创建两个表 emp_new 和 emp_old,在触发器中把 new 和 old 的内容分别添加至两个表。

```
mysql> create table emp_old like emp;
mysql> create table emp_new like emp;
mysql> delimiter /
mysql> create trigger tri_emp
    -> before update
    -> on emp
    -> for each row
    -> begin
    ->     insert into emp_new values
    ->         (new.empno, new.ename, new.job, new.mgr,
    ->          new.hiredate, new.sal, new.comm, new.deptno);
    ->     insert into emp_old values
    ->         (old.empno, old.ename, old.job, old.mgr,
    ->          old.hiredate, old.sal, old.comm, old.deptno);
    -> end
    -> /
```

对 emp 表执行 update 操作。

```
mysql> delimiter ;
mysql> update emp set sal = sal + 100 where deptno = 10;
```

查看 emp_old 和 emp_new 中的数据,注意两个查询结果中的 sal 列值的变化。

```
mysql> select * from emp_old;
+-------+--------+-----------+------+---------------------+---------+------+--------+
| empno | ename  | job       | mgr  | hiredate            | sal     | comm | deptno |
+-------+--------+-----------+------+---------------------+---------+------+--------+
|  7782 | CLARK  | MANAGER   | 7839 | 1981-06-09 00:00:00 | 2450.00 | NULL |     10 |
|  7839 | KING   | PRESIDENT | NULL | 1981-11-17 00:00:00 | 5000.00 | NULL |     10 |
|  7934 | MILLER | CLERK     | 7782 | 1982-01-23 00:00:00 | 1300.00 | NULL |     10 |
+-------+--------+-----------+------+---------------------+---------+------+--------+
3 rows in set (0.00 sec)

mysql> select * from emp_new;
+-------+-------+-----+-----+----------+-----+------+--------+
| empno | ename | job | mgr | hiredate | sal | comm | deptno |
```

```
+-------+--------+-----------+------+---------------------+---------+------+--------+
|  7782 | CLARK  | MANAGER   | 7839 | 1981-06-09 00:00:00 | 2550.00 | NULL |     10 |
|  7839 | KING   | PRESIDENT | NULL | 1981-11-17 00:00:00 | 5100.00 | NULL |     10 |
|  7934 | MILLER | CLERK     | 7782 | 1982-01-23 00:00:00 | 1400.00 | NULL |     10 |
+-------+--------+-----------+------+---------------------+---------+------+--------+
3 rows in set (0.00 sec)
```

11.4.3 模拟外键级联删除

修改外键值时,会锁住主表的相关记录,为了避免锁的产生,某些应用会限制使用外键。

下面的示例模拟外键的级联删除效果,即删除主表的记录时,级联删除子表中的相关引用记录。

```
mysql> delimiter ;
mysql> create trigger tri_del_cascade_dept
    -> after delete
    -> on dept
    -> for each row
    -> begin
    ->     delete from emp where deptno = old.deptno;
    -> end
    -> /
```

删除 emp 表的外键约束。

```
mysql> delimiter ;
mysql> alter table emp drop constraint fk_deptno;
```

在对 dept 表执行删除操作之前,查看 emp 表的行。

```
mysql> select * from emp where deptno = 10;
```

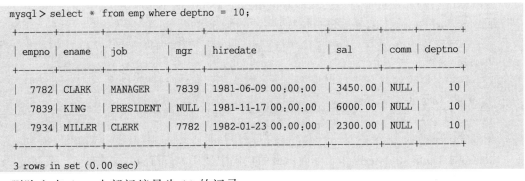

```
3 rows in set (0.00 sec)
```

删除主表 dept 中部门编号为 10 的记录。

```
mysql> delete from dept where deptno = 10;
Query OK, 1 row affected (0.01 sec)
```

确认子表 emp 中的相应记录也被删除。

```
mysql> select * from emp where deptno = 10;
Empty set (0.00 sec)
```

11.4.4 约束检查

触发器可以根据应用实际的要求,重新订正 update 操作的结果,或取消 update 操作。

下面的示例为重新订正修改结果。

触发器对工资更新前后的差值进行检查,新工资与原工资的差值不能大于1 000,若大于1 000,则把差值重设为1 000。

```
mysql> delimiter /
mysql> create trigger tri_chk_sal_diff_emp
    -> before update
    -> on emp
    -> for each row
    -> begin
    ->     if new.sal - old.sal > 1000 then
    ->         set new.sal = old.sal + 1000;
    ->     end if;
    -> end
    -> /
```

在 update 操作之前,查看 SMITH 的原工资值。

```
mysql> select ename, sal from emp where ename = 'SMITH';
+-------+--------+
| ename | sal    |
+-------+--------+
| SMITH | 800.00 |
+-------+--------+
1 row in set (0.00 sec)
```

给 SMITH 的工资增加 1 500。

```
mysql> update emp set sal = sal + 1500 where ename = 'SMITH';
```

查看修改后的新工资值,可以发现,其值增加了 1 000,而不是 1 500。

```
mysql> select ename, sal from emp where ename = 'SMITH';
+-------+---------+
| ename | sal     |
+-------+---------+
| SMITH | 1800.00 |
+-------+---------+
1 row in set (0.00 sec)
```

根据要求,取消 update 操作。下面示例的代码中使用 signal sqlstate 和 set message_text 语句指定错误号和报错信息,执行这两个语句后,update 操作也自动被取消。

修改 emp 表的工资时,不允许新值大于 8 000,若违反此要求,则取消操作,并报错。

```
mysql> delimiter /
mysql> create trigger tri_bu_emp
    -> before update on emp
    -> for each row
    -> begin
    ->     if new.sal > 8000 then
    ->         signal sqlstate 'HY000'
```

```
    ->         set message_text = 'salary cannot exceed 8000.';
    ->       end if;
    -> end
    -> /
```

执行下面的命令,把 10 号部门的工资设置为 9 500,触发器取消了此操作。

```
mysql> delimiter ;
mysql> update emp set sal = 9500 where deptno = 10;
ERROR 1644 (HY000): salary cannot exceed 8000.
mysql> select ename, sal, deptno from emp where deptno = 10;
+--------+---------+--------+
| ename  | sal     | deptno |
+--------+---------+--------+
| CLARK  | 2450.00 |     10 |
| KING   | 5000.00 |     10 |
| MILLER | 1300.00 |     10 |
+--------+---------+--------+
3 rows in set (0.00 sec)
```

11.4.5 审计

审计即把相关操作的属性记录至审计表中,以备查看,审计表需要手工创建。

下面的示例为审计 emp 表的工资修改操作。

首先创建审计表。

```
mysql> create table audit_emp
    -> (
    ->     user_id char(20),
    ->     empno int,
    ->     old_sal decimal(7,2),
    ->     new_sal decimal(7,2),
    ->     upd_time datetime
    -> );
```

user_id 存放执行 update 操作的 MySQL 用户名称,empno 存放被修改记录的员工号,old_sal 和 new_sal 存放修改前后的工资值,upd_time 存放修改操作的执行时刻。

创建触发器,比较修改前后的 sal 值,以确认 update 操作是否修改了工资值,若条件为真,则把相关属性作为一行记录加入审计表。

```
mysql> delimiter /
mysql> create trigger tri_upd_sal
    -> after update
    -> on emp
    -> for each row
    -> begin
    ->     if new.sal != old.sal then
    ->         insert into audit_emp values
    ->             (user(), new.empno, old.sal, new.sal, now());
```

```
        -> end if;
        -> end
        -> /
Query OK, 0 rows affected (0.01 sec)

mysql> delimiter ;
```

执行下面的 update 操作,修改 sal 的值。

```
mysql> update emp set sal = sal + 1000 where deptno = 10;
```

查看 audit_emp 表,可以发现以上 update 操作的相关属性已被记录。

```
mysql> select * from audit_emp;
+----------------+-------+---------+---------+---------------------+
| user_id        | empno | old_sal | new_sal | upd_time            |
+----------------+-------+---------+---------+---------------------+
| root@localhost |  7782 | 2450.00 | 3450.00 | 2020-12-05 10:52:30 |
| root@localhost |  7839 | 5000.00 | 6000.00 | 2020-12-05 10:52:30 |
| root@localhost |  7934 | 1300.00 | 2300.00 | 2020-12-05 10:52:30 |
+----------------+-------+---------+---------+---------------------+
3 rows in set (0.00 sec)
```

11.4.6 查看触发器信息

查看数据库中所有触发器的信息,最简单的方式是使用 show triggers 命令,如下面命令所示。

```
mysql> show triggers\G
```

也可以使用 from 子句指定数据库,使用 like 子句指定触发器所在的表名,表名可以使用"％"和"_"通配符,其用法与字符串模糊查询相同。

显示 db 数据库中,表名包含"emp"的触发器信息。

```
mysql> show triggers from db like '%emp%'\G
```

还可以使用 where 子句指定更多限制条件。

```
mysql> show triggers from db where `table` = 'emp' and `event` = 'UPDATE'\G
```

where 子句中包含的列名取自 show triggers 的显示结果,列名需要用符号"`"括起来,要注意,这个符号不是单引号,在键盘上,它位于数字 1 键的左侧。另外,条件中出现的字符串常量要区分大小写。

show create trigger 命令和 information_schema 系统库中的 triggers 视图,也可以用来查看触发器的系统信息,请参考下节相关内容。

11.5 查看可编程对象系统信息

查看存储程序定义可使用 show create 命令,查看更多属性可以使用 information_schema 系统库中的 routines 和 triggers 视图。

11.5.1 使用 show 命令查看程序定义

show create 命令的语法如下：

show create{procedure | function | trigger} [*db_name*].*routine_name*

查看 db 数据库中的存储过程 get_sumsal_by_dno 的定义：

```
mysql> show create procedure db.get_sumsal_by_dno\G
```

查看 db 数据库中的函数 sum_of_sal_comm 的定义：

```
mysql> show create function db.sum_of_sal_comm\G
```

查看 db 数据库中的触发器 tri_upd_sal 的定义：

```
mysql> show create trigger db.tri_upd_sal\G
```

11.5.2 使用 information_schema 系统库的 routines 和 triggers 视图

information_schema 系统库中的 routines 视图包含了存储过程和函数的系统信息，triggers 视图包含了触发器的系统信息。

routines 视图包括存储过程或函数的名称、数据库、类型、定义等信息，但不包括参数信息。

```
mysql> select routine_name, routine_type, routine_definition
    -> from information_schema.routines
    -> where routine_schema = 'db'
    -> \G
*************************** 1. row ***************************
      ROUTINE_NAME: big_table
      ROUTINE_TYPE: PROCEDURE
ROUTINE_DEFINITION: BEGIN
        DECLARE i INT DEFAULT 1;
        SET autocommit = 0;
        DROP TABLE IF EXISTS big_table;
        CREATE TABLE big_table(id INT, data VARCHAR(30));
        WHILE (i <= cnt) DO
        INSERT INTO big_table VALUES(i, CONCAT("record ", i));
            SET i = i+1;
        END WHILE;
        COMMIT;
        SET autocommit = 1;
END
1 rows in set (0.00 sec)
```

triggers 视图可以查询触发器的系统信息，包括触发器所在的表名、数据库名，以及定义（ACTION_STATEMENT 列）等属性。

```
mysql> select * from information_schema.triggers
    -> where trigger_schema = 'db'
    -> \G
*************************** 1. row ***************************
           TRIGGER_CATALOG: def
            TRIGGER_SCHEMA: db
              TRIGGER_NAME: tri_del_cascade_dept
        EVENT_MANIPULATION: DELETE
      EVENT_OBJECT_CATALOG: def
       EVENT_OBJECT_SCHEMA: db
        EVENT_OBJECT_TABLE: dept
              ACTION_ORDER: 1
          ACTION_CONDITION: NULL
          ACTION_STATEMENT: begin
    delete from emp where deptno = old.deptno;
end
        ACTION_ORIENTATION: ROW
             ACTION_TIMING: AFTER
...
1 rows in set (0.01 sec)
```

11.6 删除可编程对象

MySQL 的存储过程、函数和触发器不能使用 alter 命令修改，只能删除后重建。删除各种可编程对象的语法相似，即

drop {procedure | function | trigger} [*db_name*].*routine_name*

下面是删除存储过程、函数和触发器的实例。

```
mysql> drop procedure db.get_sumsal_by_dno;
mysql> drop function db.sum_of_sal_comm;
mysql> drop trigger db.tri_upd_sal;
```

第 12 章　服务器体系结构

运行 MySQL 服务器，需要多种组件协同工作，主要包括内存结构、系统数据库，以及重做文件等。为了适应不同的环境，MySQL 提供了服务器参数，管理员可以通过调整参数，使其高效运行。本章介绍的服务器运行在 InnoDB 存储引擎中。

本章主要内容包括：
- 总体结构及内存结构；
- 配置服务器参数；
- 事件日志文件；
- 重做文件；
- 系统数据库。

12.1　总体结构

MySQL 服务器总体结构包括磁盘和内存两部分。磁盘部分主要包括数据文件及重做文件，内存部分主要包括数据缓冲区和日志缓冲区，图 12-1 所示为 MySQL 服务器体系结构。

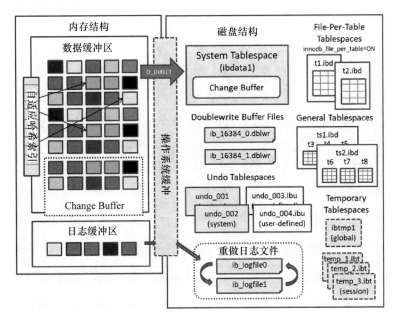

图 12-1　MySQL 服务器体系结构

12.2 内存结构

一个查询请求数据时,总是从内存的数据缓冲区读取,如果数据尚不在缓冲区,则需要从表空间读至内存。除了存放读出的数据外,数据缓冲区中还有一部分区域称为 change buffer,用于临时存放索引的修改数据。

执行数据修改时,这些修改写入日志缓冲区,也写回内存数据缓冲区(innodb buffer pool)。日志缓冲区的重做数据最后会被写入磁盘的重做日志文件。重做日志文件大小固定,最少为 2 个,循环写入,即一个文件写满后,转到另外一个继续写入。

内存数据缓冲区的数据最后会被写回表空间的数据文件。因为 MySQL 数据页一般比操作系统数据页大,为了避免在服务器发生故障时,发生部分写入的情况,MySQL 先把内存数据缓冲区的脏页写入磁盘的 doublewrite 缓冲(数据目录下的 dblwr 文件),脏页即读出后修改过的数据页。脏页写入数据文件后,其在日志文件对应的重做数据才能被覆盖重用。

12.2.1 内存数据缓冲区

内存数据缓冲区是 MySQL 各内存区的最重要部分。内存数据缓冲区的大小由系统参数 innodb_buffer_pool_size 指定(单位为字节),默认大小为 128 MB,对于生产数据库来说,此值显然过小。

```
mysql> show variables like 'innodb_buffer_pool_size';
+-------------------------+-----------+
| Variable_name           | Value     |
+-------------------------+-----------+
| innodb_buffer_pool_size | 134217728 |
+-------------------------+-----------+
1 row in set (0.00 sec)
```

数据缓冲区的大小是否合适,可以通过执行下面的查询来判定其读取操作的内存命中率。

```
mysql> select (1 - (number_pages_read / number_pages_get)) * 100
    -> from information_schema.innodb_buffer_pool_stats;
+----------------------------------------------------+
| (1 - (number_pages_read / number_pages_get)) * 100 |
+----------------------------------------------------+
|                                            93.5368 |
+----------------------------------------------------+
1 row in set (0.00 sec)
```

上面的查询结果应该尽量接近 100%,否则应该考虑增加数据缓冲区的大小。

在修改普通非唯一索引记录时,如果相应数据页不在数据缓冲区,修改结果将暂时保存在数据缓冲区的一部分内存中,这部分内存称为 change buffer。change buffer 的大小由系统参数 innodb_change_buffer_max_size 指定,默认为 25,即占整个数据缓冲区的 25%。

```
mysql> show variables like 'innodb_change_buffer_max_size';
+-------------------------------+-------+
| Variable_name                 | Value |
+-------------------------------+-------+
| innodb_change_buffer_max_size | 25    |
+-------------------------------+-------+
1 row in set (0.00 sec)
```

下次在读取磁盘上这个数据页时,读出的数据与 change buffer 内本属于这个数据页的数据合并。关闭 mysqld 服务时,change buffer 的内容会存入系统表空间的 change buffer 中。

服务器重启后,数据缓冲区内的数据都消失了,操作相关数据时,要从表空间数据文件重新读取至数据缓冲区,在重启后开始的一段时间内,服务器运行效率会因此受影响。如果重启后,立刻把之前在数据缓冲区内频繁使用的数据页恢复回来,而不是使用数据时才读取进来,会显著提高运行效率。

为此,MySQL 在关闭服务器时,会把最近使用的数据缓冲区的部分内容所在的数据页号导出至磁盘,存储在数据目录的 ib_buffer_pool 文件中。服务器再次启动时,MySQL 会根据 ib_buffer_pool 文件中保存的数据页号,把相应数据页的内容一次性地读入数据缓冲区。

系统参数 innodb_buffer_pool_dump_pct 用于设置在关闭服务器时,需要导出的数据量,默认为 25%。导出时,MySQL 会优先导出最近使用的数据页。

查看 innodb_buffer_pool_dump_pct 参数值,以及数据目录的 ib_buffer_pool 文件。

```
mysql> show variables like 'innodb_buffer_pool_dump_pct';
+-----------------------------+-------+
| Variable_name               | Value |
+-----------------------------+-------+
| innodb_buffer_pool_dump_pct | 25    |
+-----------------------------+-------+
1 row in set (0.00 sec)
[root /var/lib/mysql 2020-09-27 10:15:24]
# ls -lh ib_buffer*
-rw-r-----. 1 mysql mysql 4.1K Sep 26 21:56 ib_buffer_pool
```

12.2.2 内存日志缓冲区

内存日志缓冲区用来缓存重做数据。系统参数 innodb_log_buffer_size 用来设置其大小,默认为 16 MB。当日志缓冲区写满,或事务提交时,MySQL 会把日志缓冲区的数据写入磁盘的重做文件。

```
mysql> show variables like 'innodb_log_buffer_size';
+------------------------+----------+
| Variable_name          | Value    |
+------------------------+----------+
| innodb_log_buffer_size | 16777216 |
+------------------------+----------+
1 row in set (0.00 sec)
```

12.2.3 排序缓冲区和连接缓冲区

用户执行 order by、group by 等需要排序的查询操作时,MySQL 对其分配排序缓冲区。排序缓冲区的大小由 sort_buffer_size 指定,默认为 256 KB。

用户执行表连接查询时,如果连接字段没有索引,MySQL 需要执行全表扫描读取两个表的行,然后在连接缓冲区对读出的记录执行连接操作。连接缓冲区的大小由 join_buffer_size 指定,默认为 256 KB。

若以上相关操作的数据量大,可以适当增大当前会话的 sort_buffer_size 和 join_buffer_size 参数值。

12.2.4 内部临时表内存

用户执行子查询、union 等操作时,MySQL 在内存中使用内部临时表实现。临时表使用 TempTable 引擎创建,用于存放临时表的内存大小由参数 temptable_max_ram 指定,默认为 1 GB,这部分内存由各用户共享。当使用的内存超出这个限制时,内部临时表会映射为磁盘的临时文件。

12.2.5 内存的自动设置

从 MySQL 8.0.14 开始,如果服务器硬件只作为 MySQL 数据库服务器运行 mysqld 服务,那么可以开启系统参数 innodb_dedicated_server。此参数开启后,MySQL 会根据物理内存的大小自动配置内存数据缓冲区大小以及重做日志相关参数。

innodb_dedicated_server 为静态参数,默认关闭。需要在/etc/my.cnf 中配置为开启,重启服务器才生效。

```
mysql> show variables like 'innodb_dedicated_server';
+-------------------------+-------+
| Variable_name           | Value |
+-------------------------+-------+
| innodb_dedicated_server | OFF   |
+-------------------------+-------+
1 row in set (0.00 sec)
```

MySQL 会根据物理内存的大小自动配置内存参数 innodb_buffer_pool_size,如表 12-1 所示。

表 12-1 内存数据缓冲区的自动配置

服务器物理内存(pm)	innodb_buffer_pool_size
pm ≤ 1 GB	128 MB
1 GB ≤ pm ≤ 4 GB	pm × 0.5
pm > 4 GB	pm × 0.75

12.3 配置服务器和客户端参数

在服务器启动时,MySQL 需要确定如何配置各内存结构的大小,以及需要操作的各种文件的位置、大小等属性,这些属性要在启动 mysqld 时,以参数形式给出。

客户端连接服务器时,要指定服务器地址、登录服务器所用的用户名以及密码,有时为了使用某种特点的显示环境,也需要指定客户端环境参数。

12.3.1 设置方式

设置 mysqld 服务参数,可以通过以下 3 种方式。
- 命令行参数;
- 配置文件;
- set 命令。

命令行参数方式是指在启动 mysqld 服务的同时,指定相关参数值。在服务器运行期间,这些参数值保持有效,但服务器重启后会失效。

MySQL 的默认配置文件为/etc/my.cnf,它是一个文本文件,分为若干以方括号标识的参数组,组里的一行表示一个参数值。修改配置文件中的参数时,若参数存在,则直接修改其值;若参数不存在,则在此文件的相应参数组中增加一行。

[mysqld]组中的参数用于设置服务器,启动 mysqld 时,会读取其中的内容。

[client]组为公共组,启动 mysql、mysqldump、mysqlbinlog 等客户端工具时,会读取此组中的参数。

除读取[client]组外,各客户端还会读取与客户端名称相同的参数组,如启动 mysql 时,也会读取[mysql]组,如果一个参数在多个组出现,则后面的设置会覆盖前面的设置。

set 命令设置系统参数时可以使用 global、session、persist 以及 persist_only 选项,使得设置结果有不同的生效范围和生命周期。

多数参数可以使用命令行参数、配置文件以及 set 命令这 3 种方式的任意一种修改。个别参数只支持命令行或配置文件,如用于指定默认配置文件的 defaults-file、输出配置文件的参数值 print-defaults 等只能以命令行方式指定,而 innodb_dedicated_server 参数只能在配置文件中设置。

12.3.2 使用命令行参数

在启动 mysqld 时,指定参数。

```
# mysqld -- user = mysql -- innodb-buffer-pool-size = 512M &
```

注意"="两侧不能留有空格,另外参数名称中的分隔符可以使用"-"或"_",习惯上使用"-"。

通过以上方式启动的 mysqld,可以在 mysql 中执行 shutdown 命令将其关闭。

```
mysql > shutdown;
Query OK, 0 rows affected (0.00 sec)
```

12.3.3 使用配置文件

使用配置文件设置 innodb_buffer_pool_size,只要把下面一行加入[mysqld]组内即可。

```
[mysqld]
innodb_buffer_pool_size = 512M
```

参数名称中的分隔符可以使用"-"或"_",innodb_buffer_pool_size 也可以写为 innodb_

buffer_pool-size,习惯上使用"_"。

12.3.4 使用 set 命令

使用 set 命令设置系统参数时,可以使用下面几个选项。
- set global;
- set session;
- set persist;
- set persist_only。

多数参数可以使用选项 global、persist,其共同特点是可以修改当前运行值,并在整个服务器范围生效,而不需要重启服务器,这种参数称为动态参数。使用 global 选项修改的参数在服务器重启后失效,使用 persist 选项修改的参数会保存至数据目录的 mysqld-auto.cnf 文件,服务器重启后依然生效。mysqld-auto.cnf 文件为 JSON 格式。

使用 session 选项只对当前会话生效,退出连接即失效。部分参数的修改对于 global 和 session 选项都支持,部分参数只支持其一,如 innodb_buffer_pool_size 只支持 global 选项修改,而 sql_log_bin 只支持 session 选项修改。

set 命令的 persist_only 选项不会修改运行值,只把新值存入数据目录的 mysqld-auto.cnf 文件。若参数不允许修改运行值(即静态参数),除了可以使用命令行参数或配置文件设置参数值外,一般也可以使用 set 命令的 persist_only 选项。

下面分别使用以上几种选项设置 innodb_buffer_pool_size。

使用 global 选项修改。

```
mysql> set global innodb_buffer_pool_size = 268435456;
```

innodb_buffer_pool_size 不支持使用 session 选项修改。

```
mysql> set session innodb_buffer_pool_size = 268435456;
ERROR 1229 (HY000): Variable 'innodb_buffer_pool_size' is a GLOBAL variable and should be set with SET GLOBAL
```

使用 persist 选项修改。

```
mysql> set persist innodb_buffer_pool_size = 268435456;
```

使用 persist_only 选项修改。

```
mysql> set persist_only innodb_buffer_pool_size = 268435456;
```

要把参数设置为默认值,指定为 default 即可。

```
mysql> set global innodb_buffer_pool_size = default;
```

从 mysqld-auto.cnf 中除去某个参数,执行 reset persist。

```
mysql> reset persist innodb_buffer_pool_size;
```

12.3.5 查看系统参数

使用 show session variables 命令,查看当前会话生效的参数值,session 可省略。

```
mysql> show session variables\G
*************************** 1. row ***************************
Variable_name: activate_all_roles_on_login
        Value: ON
```

```
*************************** 2. row ***************************
Variable_name: admin_address
        Value:
...
*************************** 608. row ***************************
Variable_name: windowing_use_high_precision
        Value: ON
608 rows in set (0.01 sec)
```

使用 show global variables 命令,查询服务器范围生效的参数值。

```
mysql> show global variables\G
*************************** 1. row ***************************
Variable_name: activate_all_roles_on_login
        Value: OFF
*************************** 2. row ***************************
Variable_name: admin_address
        Value:
...
*************************** 586. row ***************************
Variable_name: windowing_use_high_precision
        Value: ON
586 rows in set (0.00 sec)
```

使用 like 关键字,查询指定参数的值。

```
mysql> show variables like 'innodb_buffer_pool_size';
+-------------------------+-----------+
| Variable_name           | Value     |
+-------------------------+-----------+
| innodb_buffer_pool_size | 134217728 |
+-------------------------+-----------+
1 row in set (0.00 sec)
```

使用通配符%,查询参数名称以 innodb_buffer_pool 开头的参数的值。

```
mysql> show variables like 'innodb_buffer_pool%';
+-------------------------------------+-----------+
| Variable_name                       | Value     |
+-------------------------------------+-----------+
| innodb_buffer_pool_chunk_size       | 134217728 |
| innodb_buffer_pool_dump_at_shutdown | ON        |
...
| innodb_buffer_pool_size             | 134217728 |
+-------------------------------------+-----------+
11 rows in set (0.00 sec)
```

对于同一参数,其 global 和 session 级别的值可能不同,设置和查询参数值时,注意附加相应选项。

```
mysql> set session transaction_isolation = 'read-uncommitted';
Query OK, 0 rows affected (0.00 sec)

mysql> set global transaction_isolation = 'read-committed';
Query OK, 0 rows affected (0.00 sec)

mysql> show session variables like 'transaction_isolation';
+-----------------------+------------------+
| Variable_name         | Value            |
+-----------------------+------------------+
| transaction_isolation | READ-UNCOMMITTED |
+-----------------------+------------------+
1 row in set (0.00 sec)

mysql> show global variables like 'transaction_isolation';
+-----------------------+----------------+
| Variable_name         | Value          |
+-----------------------+----------------+
| transaction_isolation | READ-COMMITTED |
+-----------------------+----------------+
1 row in set (0.00 sec)
```

mysqld-auto.cnf 文件的格式为 JSON，不方便直接查看，performance_schema 系统库的 persisted_variables 视图，可用来查看 mysqld-auto.cnf 文件中的参数。

```
mysql> select * from performance_schema.persisted_variables;
+----------------------------+----------------+
| VARIABLE_NAME              | VARIABLE_VALUE |
+----------------------------+----------------+
| password_history           | 5              |
| password_reuse_interval    | 60             |
| innodb_undo_log_truncate   | ON             |
| password_require_current   | ON             |
| default_password_lifetime  | 90             |
+----------------------------+----------------+
5 rows in set (0.00 sec)
```

12.4 事件日志文件

事件日志文件都是文本文件，主要包括错误日志、通用查询日志及慢查询日志。用户或管理员可以利用这些文件查看 MySQL 的运行状态。

12.4.1 设置日志的时区和输出目标

两个系统参数可以影响事件日志文件。log_timestamps 参数用于设置日志文件中的时区，其影响范围包括错误日志、通用查询日志及慢查询日志。log_output 参数设置通用查询日志和慢查询日志的输出目标。

事件日志的时区属性由系统参数 log_timestamps 指定,默认为 UTC,与北京时间相差 8 个小时。若要设置为服务器本地系统时间,可以将其修改为 system。

```
mysql> set persist log_timestamps = system;
```

系统参数 log_output 的可选值为 FILE、TABLE 或 NONE,默认值为 FILE,即通用查询日志和慢查询日志输出为文本文件,指定为 NONE 时,不输出任何信息。

查询 log_output 的当前值。

```
mysql> show variables like 'log_output';
+---------------+-------+
| Variable_name | Value |
+---------------+-------+
| log_output    | FILE  |
+---------------+-------+
1 row in set (0.00 sec)
```

若将 log_output 设置为 TABLE,则通用查询日志以及慢查询日志信息会分别输出至 mysql 系统库的 general_log 表及 slow_log 表,从而方便用户指定列和查询条件,并可以远程访问。这两个表使用 CSV 存储引擎,存储格式为逗号分隔的文本文件。

也可以将 log_output 设置为 FILE 和 TABLE,每种格式输出一份。

把 log_output 设置为 FILE 和 TABLE。

```
mysql> set persist log_output = 'file,table';
```

12.4.2 错误日志

错误日志文件存储 MySQL 服务器启动、关闭过程以及运行状态信息。服务器发生故障时,管理员可以通过检查错误日志文件,查找故障原因,或发现潜在问题。

系统参数 log_error 用于指定错误日志文件的目录和名称,默认值为/var/log/mysqld.log。

```
mysql> show variables like 'log_error';
+---------------+---------------------+
| Variable_name | Value               |
+---------------+---------------------+
| log_error     | /var/log/mysqld.log |
+---------------+---------------------+
1 row in set (0.00 sec)
```

log_error 为静态参数,要在/etc/my.cnf 中修改其值,服务器重启后生效。

```
[mysqld]
log_error = /var/log/mysqld.log
```

从 MySQL 8.0.22 开始,错误日志也会写入 performance_schema.error_log 表,查看错误日志时,可以不登录操作系统,也可以附加条件。

下面是查询 error_log 表的示例。

```
mysql> select * from performance_schema.error_log
    -> where data like '% temporary password %'\G
*************************** 1. row ***************************
   LOGGED: 2021-03-05 08:24:06.512736
THREAD_ID: 6
     PRIO: Note
ERROR_CODE: MY-010454
SUBSYSTEM: Server
     DATA: A temporary password is generated for root@localhost: f!BUk9m(># ik
1 row in set (0.00 sec)
```

12.4.3 通用查询日志

通用查询日志(general query log)会记录 mysqld 执行的任务,会记录客户端的连接和断开,也会记录客户端发送到服务器端的每条 SQL 命令。

若系统参数 binlog_format 设置为 row,则在执行 update 操作时,并不发送 SQL 语句,而是发送改变的数据,从而 update 操作不会记录至查询日志。当系统参数 binlog_format 设置为 miexed 时,update 语句也可能不记录至查询日志。

geneal_log 系统参数用来设置是否开启通用查询日志功能,默认为 off,即关闭。

```
mysql> show variables like 'general_log';
+---------------+-------+
| Variable_name | Value |
+---------------+-------+
| general_log   | OFF   |
+---------------+-------+
1 row in set (0.00 sec)
```

开启通用查询日志功能后,日志文件名称由系统参数 general_log_file 指定,默认在数据目录下,名称为 *hostname*.log

```
mysql> show variables like 'general_log_file';
+------------------+---------------------+
| Variable_name    | Value               |
+------------------+---------------------+
| general_log_file | /var/lib/mysql/ol83.log |
+------------------+---------------------+
1 row in set (0.01 sec)
```

执行下面几个简单命令,验证产生的通用日志内容。

```
# mysql
mysql> use db
mysql> select * from dept;
mysql> delete from emp where empno = 8888;
mysql> exit
```

显示通用查询日志文件的内容,可以看到上面的命令对应的日志记录。

```
[root /var/lib/mysql 2021-03-01 22:08:35]
# cat ol83.log
/usr/sbin/mysqld, Version: 8.0.23 (MySQL Community Server - GPL). started with:
Tcp port: 3306  Unix socket: /var/lib/mysql/mysql.sock
Time                        Id Command     Argument
2021-03-01T14:07:46.637715Z  8 Query       show variables like 'general_log'
2021-03-01T14:08:02.814528Z  8 Query       SELECT DATABASE()
2021-03-01T14:08:02.814993Z  8 Init DB     db
2021-03-01T14:08:02.816392Z  8 Query       show databases
2021-03-01T14:08:02.821393Z  8 Query       show tables
2021-03-01T14:08:02.824533Z  8 Field List  b1
2021-03-01T14:08:02.898747Z  8 Field List  b2
2021-03-01T14:08:02.901571Z  8 Field List  dept
2021-03-01T14:08:02.903213Z  8 Field List  emp
2021-03-01T14:08:02.903687Z  8 Field List  salarygrade
2021-03-01T14:08:02.904566Z  8 Query       select * from dept
2021-03-01T14:08:02.906733Z  8 Query       delete from emp where empno = 8888
2021-03-01T14:08:02.911908Z  8 Quit
```

12.4.4 慢查询日志

慢查询日志用于确定执行时间过长的 SQL 命令，其设置与使用方法与通用查询日志类似。下面几个系统参数用于设置慢查询日志的各种属性。

slow_query_log：设置慢查询日志功能是否开启，默认为 off，即关闭。

slow_query_log_file：设置日志文件的名称，默认为数据目录下的 *hostname*-slow.log。

long_query_time：指定慢查询的时间阈值，单位是 s，默认为 10，即执行时间超过 10 s 的 SQL 命令会作为慢查询记录。

12.5 重做文件

重做文件存储用户对 MySQL 服务器的操作记录，如建表、数据修改等操作。读取数据的操作，如查询或 show 命令不会记入重做文件。数据库发生故障时，重做文件用于恢复数据库。MySQL 的重做文件包括两种：二进制日志文件和重做日志文件，英文名称分别为 binary log 和 redo log，前者属于整个服务器，用于存储介质发生故障后的数据库恢复，后者只用于 InnoDB 存储引擎中的实例恢复（由停电、死机等情况造成）。

12.5.1 binary log

binary log 存储用户执行的命令，也称逻辑形式。

在服务器复制环境下，binary log 从主服务器传送至从服务器，然后在从服务器重新执行，以达到主从服务器数据同步的目的。binary log 用于 MySQL 服务器的所有存储引擎。

使用数据库备份恢复数据库时，把之前的备份还原后，继续使用备份操作后产生的 binary log，把数据库恢复至出现故障的时刻，这种恢复也称为增量恢复（incremental recovery）。

MySQL 默认在数据目录创建 binary log 文件,其文件名为 binlog,扩展名为从 000001 开始的 6 位数字。MySQL 每次创建 binary log 时,新文件的文件名不变,扩展名在之前的基础上加 1。数据目录下也会创建一个索引文件 binlog.index,记录服务器正在使用和使用过的 binary log 文件全名。

binary log 的主要属性可以通过下面几个系统参数设置或查看。

disable_log_bin:关闭 binary log。作为一行添加在/etc/my.cnf 中,重启 mysqld 生效。

log-bin:设置 binary log 的路径和文件基名(basename,即不包括扩展名),需要在/etc/my.cnf 中设置,重启 mysqld 生效。索引文件也会自动创建在 log-bin 指定的目录下。

max_binlog_size:设置 binary log 的大小上限,可以使用 set persist 命令设置。

log_bin:只读参数,用于查看 binary log 是否开启,其值为 on 或 off,取决于/etc/my.cnf 中是否添加了 disable_log_bin。

log_bin_basename:只读参数,用于查看 binary log 的文件基名,即 log-bin 的值。

sql_log_bin:设置当前会话是否关闭 binary log,使用 set 命令执行。

下面用实例说明以上参数的用法。

若需要关闭 binary log,可以在/etc/my.cnf 中添加一行 disable_log_bin。去除此行即开启 binary log。

```
[mysqld]
disable_log_bin
```

重启 mysqld 后,可以确认系统参数 log_bin 的值为 off。

```
mysql> show variables like 'log_bin';
+---------------+-------+
| Variable_name | Value |
+---------------+-------+
| log_bin       | OFF   |
+---------------+-------+
1 row in set (0.06 sec)
```

在/var/lib 创建子目录 mybinlog,并指定其属主和属组为 mysql:mysql

```
[root /var/lib 2021-02-23 09:23:50]
# mkdir mybinlog
[root /var/lib 2021-02-23 09:25:01]
# chown -R mysql:mysql mybinlog
```

在/etc/my.cnf 中去除 disable_log_bin,添加下面一行设置 binary log 的目录和文件基名。

```
[mysqld]
log-bin = /var/lib/mybinlog/binary_log
```

重启 mysqld 后,查看/var/lib/mybinlog 目录下的内容。

```
[root /var/lib/mybinlog 2021-02-23 09:26:29]
# ls -l
total 8
-rw-r-----. 1 mysql mysql 156 Feb 23 09:26 binary_log.000001
-rw-r-----. 1 mysql mysql  36 Feb 23 09:26 binary_log.index
```

查看 log_bin_basename 的当前值,可以确认已赋予 log-bin 的值。

```
mysql> show variables like 'log_bin_basename';
+------------------+-----------------------------+
| Variable_name    | Value                       |
+------------------+-----------------------------+
| log_bin_basename | /var/lib/mybinlog/binary_log |
+------------------+-----------------------------+
1 row in set (0.00 sec)
```

系统参数 max_binlog_size 用于设置 binary log 文件的最大大小,默认为 1 GB,这也是可设置范围的最大值。

```
mysql> show variables like 'max_binlog_size';
+-----------------+------------+
| Variable_name   | Value      |
+-----------------+------------+
| max_binlog_size | 1073741824 |
+-----------------+------------+
1 row in set (0.00 sec)
```

修改 max_binlog_size 的值为 512 MB。

```
mysql> set persist max_binlog_size = 536870912;
```

一个事务产生的 binary log 不能跨文件存储,如果执行大的事务,binary log 文件的大小可能会超过 max_binlog_size 的设置值。

系统参数 sql_log_bin 用于设置当前会话是否记录 binary log,若为 off,则当前会话执行的 SQL 命令不会存入 binary log。

```
mysql> set sql_log_bin = off;
```

下面将介绍 binary log 的常见操作。

在服务器重启或文件大小达到 max_binlog_size 设置的上限时,MySQL 才会自动创建和使用新的 binary log 文件。若需要手工创建新的 binary log 文件,可以执行下面的命令。

```
mysql> flush binary logs;
```

执行数据库备份后,备份操作之前产生的 binary log 文件就不再需要了,可以将其删除。

下面的命令把最后写入时间在'2021-02-23 10:05:00'之前的 binary log 文件都删除。

```
mysql> purge binary logs before '2021-02-23 10:05:00';
```

下面的命令把扩展名序号小于 binary_log.000005 的 binary log 文件都删除。

```
mysql> purge binary logs to 'binary_log.000005';
```

相比手工删除,更有效的方法是通过指定系统参数 binlog_expire_logs_seconds 的值,设置 binary log 文件的有效期,让 MySQL 自动删除。

binlog_expire_logs_seconds 的默认值为 2 592 000 s,即 30 天。

```
mysql> show variables like 'binlog_expire_logs_seconds';
+----------------------------+---------+
| Variable_name              | Value   |
+----------------------------+---------+
| binlog_expire_logs_seconds | 2592000 |
```

```
+----------------------------+------+
1 row in set (0.00 sec)
```

如果数据库备份周期为 7 天,则可以将其设置为 8 天,即 691 200 s。

```
mysql> set persist binlog_expire_logs_seconds = 691200;
```

服务器重启或执行 flush binary logs 命令时,最后写入时间在 binlog_expire_logs_seconds 设定值之前的 binary log 文件会被自动删除。

查看正在使用的 binary log 文件及当前数据量。

```
mysql> show master status;
+-------------------+----------+
| File              | Position |
+-------------------+----------+
| binary_log.000007 |      156 |
+-------------------+----------+
1 row in set (0.01 sec)
```

以上的 Position 指当前的日志数据量为 156 字节,即日志文件的第 0 字节至第 155 字节已经写入数据,新的日志数据写入 binary log 文件时,由第 156 字节处开始写起。

执行 show binary logs 命令,查看所有的 binary log。

```
mysql> show binary logs;
+---------------+-----------+-----------+
| Log_name      | File_size | Encrypted |
+---------------+-----------+-----------+
| binlog.000001 |      3789 | No        |
| binlog.000002 |       179 | No        |
...
| binlog.000007 |       156 | No        |
+---------------+-----------+-----------+
7 rows in set (0.02 sec)
```

执行 show binlog events 命令,查看指定日志文件中的事件,from … limit 子句用于设置起始位置和事件个数。

```
mysql> show binlog events in 'binlog.000011' from 156 limit 3\G
*************************** 1. row ***************************
   Log_name: binlog.000011
        Pos: 156
 Event_type: Anonymous_Gtid
  Server_id: 1
End_log_pos: 233
       Info: SET @@SESSION.GTID_NEXT = 'ANONYMOUS'
*************************** 2. row ***************************
   Log_name: binlog.000011
        Pos: 233
 Event_type: Query
  Server_id: 1
```

```
      End_log_pos: 347
             Info: use `db`; create index idx_sal on emp(sal) /* xid = 16 */
*************************** 3. row ***************************
         Log_name: binlog.000011
              Pos: 347
       Event_type: Anonymous_Gtid
        Server_id: 1
      End_log_pos: 424
             Info: SET @@SESSION.GTID_NEXT = 'ANONYMOUS'
3 rows in set (0.00 sec)
```

另外，mysqlbinlog 工具可以把指定二进制日志文件转换为 SQL 语句构成的脚本文件，具体用法，请参考第 19 章备份恢复相关内容。

12.5.2 redo log

redo log 存储因用户操作导致的数据页改变，这种日志记录方式也称物理形式。redo log 只用于 InnoDB 存储引擎。

服务器出现死机或停电等问题时，MySQL 使用 redo log 自动恢复数据库，这种情况不使用数据库备份，也不需要用户参与。

MySQL 在初始化时，默认创建两个 redo log 文件，ib_logfile0 和 ib_logfile1，MySQL 对这两个文件循环写入重做数据。

```
[root /var/lib/mysql 2021-03-02 08:35:51]
# ls -lh ib_log*
-rw-r-----. 1 mysql mysql 48M Mar  2 07:51 ib_logfile0
-rw-r-----. 1 mysql mysql 48M Feb 25 07:27 ib_logfile1
```

改变重做日志文件的大小和个数，可以配置下面两个系统参数：innodb_log_file_size，innodb_log_files_in_group，两个参数为静态参数，需要在配置文件 my.cnf 中修改，并重启服务器使其生效。

在 MySQL 8.0.14 之前版本（或未开启 innodb_dedicated_server 参数的 8.0.13 之后版本），innodb_log_files_in_group 的值默认为 2，innodb_log_file_size 会依照服务器物理内存大小自动配置。

表 12-2　重做日志文件的自动配置

服务器物理内存(pm)	innodb_log_file_size
pm < 1 GB	48 MB
1 GB ≤ pm ≤ 4 GB	128 MB
4 GB < pm ≤ 8 GB	512 MB
8 GB < pm ≤ 16 GB	1 GB
pm > 16 GB	2 GB

若手工配置，则可以修改/etc/my.cnf。如将重做文件大小改为 100 MB，个数改为 3，可以在 my.cnf 中添加以下行。

```
[mysqld]
innodb_log_file_size = 104857600
innodb_log_files_in_group = 3
```

重启服务器后,查看重做文件的情况。

```
[root /var/lib/mysql 2021-03-02 08:37:21]
# ls -lh ib_log*
-rw-r-----. 1 mysql mysql 100M Mar  2 08:37 ib_logfile0
-rw-r-----. 1 mysql mysql 100M Mar  2 08:37 ib_logfile1
-rw-r-----. 1 mysql mysql 100M Mar  2 08:37 ib_logfile2
```

从 MySQL 8.0.14 开始,若 innodb_dedicated_server 参数设置为开启,则 innodb_log_file_size 和 innodb_log_files_in_group(最小为 2)会根据内存数据缓冲区大小(innodb_buffer_pool_size)依照如表 12-3 所示的规则自动配置。

表 12-3　重做日志文件的自动配置(8.0.14 版本及之后)

innodb_buffer_pool_size(bp)	innodb_log_file_size	innodb_log_files_in_group
bp < 8 GB	512 MB	round(bp)
8 GB ≤ bp ≤ 128 GB	1 GB	round(bp × 0.75)
bp > 128 GB	2 GB	64

12.6　系统数据库

系统数据库用来保存系统数据,以确保服务器正常运行,也可供用户查询,以了解服务器、数据库以及各种对象的系统信息。

MySQL 包括 4 个系统数据库:mysql、sys、performance_schema、information_schema。

12.6.1　mysql

mysql 数据库存储服务器上各数据库及数据库内各对象的数据字典表、用户权限系统表、帮助信息系统表、时区系统表等。

mysql 数据库的数据存储在数据目录下的 mysql.ibd 文件中,数据目录内的 mysql 子目录只存储通用查询日志和慢查询日志表。

mysql 数据库中的数据字典表存储字符集及排序规则、服务器中的所有数据库信息、各数据库内的对象信息。典型的数据字典表包括 character_sets、schemata、tablespaces、tables、indexes、routines、check_constraints 等。在 MySQL 中,schema 与 database 为同义词,schemata 为 schema 的复数形式,schemata 字典表存储所有数据库的信息。

mysql 数据库中的数据字典表对用户不可见,不会出现在 show tables 的结果中,也不会出现在 information_schema.tables 中,但可以通过 show 命令或 information_schema 数据库中的相关视图间接访问(一般与数据字典表同名)。

如果访问数据字典表,会有以下报错。

```
mysql> select * from mysql.tables;
ERROR 3554 (HY000): Access to data dictionary table 'mysql.tables' is rejected.
```

用户和权限相关系统表是 mysql 系统数据库最常用的部分,包括存储用户属性与全局权限的 user 表、存储数据库权限的 db 表、存储对象权限的 tables_priv 表等。

mysql 数据库中的时区相关表包括 time_zone、time_zone_name 等,这些表默认为空表,使用这些表时,需要先执行下面的命令加载时区数据。

```
# mysql_tzinfo_to_sql /usr/share/zoneinfo | mysql -uroot -p mysql
```

12.6.2 sys

sys 数据库由 MySQL 5.6 开始引入,主要是对 performance_schema 数据库的数据再汇总,提供另一种访问 performance_schema 数据库的方法,使其更便于查看。

sys 数据库也包含一些用于系统性能调试的系统视图和存储过程、函数。

sys 数据库内的视图往往成对出现,一个在名称前面加上 x$,其数据未经可读性加工。下面两个查询,可以看出其结果的明显区别。

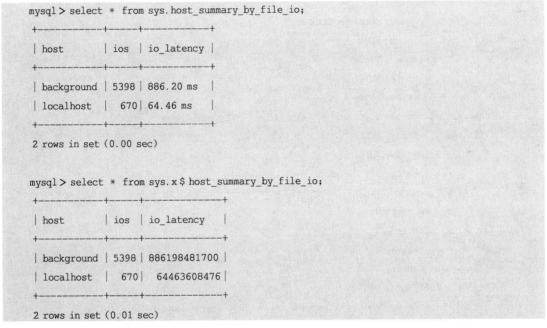

下面是 sys 数据库中的几个常用视图。

schema_object_overview 视图列出各个数据库中的对象个数统计信息。

```
| sys                        | FUNCTION       |  22 |
| sys                        | INDEX (BTREE)  |   1 |
| sys                        | PROCEDURE      |  26 |
| sys                        | TRIGGER        |   2 |
| sys                        | VIEW           | 100 |
+----------------------------+----------------+-----+
12 rows in set (0.05 sec)
```

session 视图可以查看各个连接正在执行的任务情况,这些连接不包括后台线程。

```
mysql> select * from sys.session
    -> where conn_id != connection_id()
    -> and trx_state = 'active'\G
*************************** 1. row ***************************
                thd_id: 51
               conn_id: 11
                  user: root@localhost
                    db: NULL
               command: Sleep
                 state: NULL
                  time: 121
     current_statement: update db.emp set sal = 3000 where ename = 'FORD'
     statement_latency: NULL
              progress: NULL
          lock_latency: 558.00 us
         rows_examined: 12
             rows_sent: 0
         rows_affected: 0
            tmp_tables: 0
       tmp_disk_tables: 0
             full_scan: NO
        last_statement: update db.emp set sal = 3000 where ename = 'FORD'
last_statement_latency: 10.89 ms
        current_memory: 71.58 KiB
             last_wait: NULL
     last_wait_latency: NULL
                source: NULL
           trx_latency: 2.53 min
             trx_state: ACTIVE
        trx_autocommit: NO
                   pid: 1639
          program_name: mysql
1 rows in set (0.03 sec)
```

processlist 可以查看包括后台线程在内的所有线程的情况。

schema_tables_with_full_table_scans 视图可以查看对哪些表使用了全表扫描。

```
mysql> select * from sys.schema_tables_with_full_table_scans;
+---------------+-------------+-------------------+---------+
| object_schema | object_name | rows_full_scanned | latency |
+---------------+-------------+-------------------+---------+
| db            | emp         |                13 | 9.99 ms |
| db            | dept        |                 4 | 9.27 ms |
+---------------+-------------+-------------------+---------+
2 rows in set (0.00 sec)
```

statement_analysis 视图可以查看所有 SQL 命令执行的统计情况。

statements_with_sorting 视图可以查看使用了排序操作的 SQL 命令的执行情况。

sys 数据库中也提供了一些完成常见任务的存储过程和函数。有些数据库名称比较长，如 performance_schema，使用存储过程 sys.create_synonym_db() 可以对数据库创建同义词，方便引用。

下面的示例为对 db 数据库创建同义词 db_syn。

先查看 db 数据库中表的情况。

```
mysql> show full tables from db;
+---------------+------------+
| Tables_in_db  | Table_type |
+---------------+------------+
| dept          | BASE TABLE |
| emp           | BASE TABLE |
| salgrade      | BASE TABLE |
+---------------+------------+
3 rows in set (0.01 sec)
```

对 db 数据库创建同义词 db_syn。

```
mysql> call sys.create_synonym_db('db','db_syn');
+-------------------------------------------+
| summary                                   |
+-------------------------------------------+
| Created 3 views in the `db_syn` database  |
+-------------------------------------------+
1 row in set (0.02 sec)

Query OK, 0 rows affected (0.02 sec)
```

由以上提示信息可知，创建数据库同义词，实质是创建了一个以同义词命名的新数据库，并在新数据库中创建了与源数据库的表同名的视图。这可以由下面的查询验证。

```
mysql> show full tables from db_syn;
+------------------+------------+
| Tables_in_db_syn | Table_type |
+------------------+------------+
| dept             | VIEW       |
| emp              | VIEW       |
| salgrade         | VIEW       |
```

```
+--------------------+------------+
3 rows in set (0.00 sec)

mysql> show create view db_syn.dept\G
*************************** 1. row ***************************
                View: dept
         Create View: CREATE ALGORITHM = UNDEFINED DEFINER = `root`@`localhost` SQL SECURITY
INVOKER VIEW `db_syn`.`dept` AS select `db`.`dept`.`deptno` AS `deptno`,`db`.`dept`.`dname` AS `dname`,`db`.
`dept`.`loc` AS `loc` from `db`.`dept`
   character_set_client: utf8mb4
   collation_connection: utf8mb4_0900_ai_ci
1 row in set (0.00 sec)
```

源数据库中的表删除后,同义词数据库中的同名视图不会同步删除,需要用户手工删除。对数据库创建同义词后,会在数据目录创建与同义词同名的空目录。

12.6.3 performance_schema

performance_schema 由 MySQL 5.5 引入,在底层监控 MySQL 服务器的性能。其表为内存表,只在关闭服务器时导出到磁盘,在启动时导入内存。

performance_schema 数据库中的几个常用表包括查询锁信息的 data_lock_waits 和 data_locks 表,存储全局和会话参数值的 global_variables 和 session_variables 表,以及存储全局和会话状态的 global_status 和 session_status 表,相关 show 命令的数据即来自于此。

12.6.4 information_schema

information_schema 由 MySQL 5.0 引入,其目的是要符合 SQL:2003 标准的要求,提供给用户查询各数据库以及各种对象的系统信息。此数据库不存储数据,其对象都是视图,数据来自 mysql 系统库中的数据字典基表,在数据目录中,没有对应此数据库的目录和数据文件。视图的命名及结构遵从 SQL:2003,MySQL 扩展的列在说明文档中标记为"MySQL Extension"。任何用户都可以访问此数据库,但只能查看自己有权限操作的内容。很多 show 命令即对此数据库的查询。

第 13 章 表空间和数据文件

表空间是 MySQL 的逻辑存储空间,由数据文件构成,用于存储表和索引的数据。
本章主要内容包括:
- 表空间的概念及分类;
- 系统表空间、临时表空间、undo 表空间的功能;
- file_per_table 和 general 表空间;
- 查询表和表空间的属性;
- 在表空间之间转移表。

13.1 表空间的概念

表空间可以看作放置表的容器,由数据文件构成。

13.1.1 数据目录

MySQL 的数据库数据默认存储在数据目录中,数据目录的位置由系统参数 datadir 指定,默认为/var/lib/mysql。

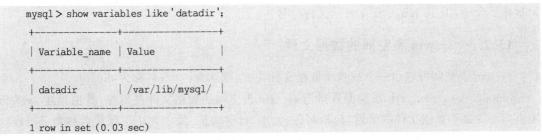

创建数据库时,会在数据目录创建同名的子目录。

13.1.2 MySQL 的表空间分类

MySQL 的表空间主要有以下几种类型:
- system;
- temporary;
- undo;
- file-per-table;
- general。

system、temporary、undo 三类表空间主要用于存放系统数据。file_per_table 和 general 表空间用于存放用户的表或索引数据。

13.2　system 表空间

系统表空间(system 表空间)用于存储 change buffer 数据。若 innodb_file_per_table 系统参数设置为 0(默认为 1),则在创建表或索引时,默认存储于 system 表空间。在 MySQL 8.0.20 版本之前,还存储 doublewrite buffer 数据。

13.2.1　change buffer

普通索引记录被修改时,如果相应数据页不在数据缓冲区,修改结果暂时保存在数据缓冲区的一部分内存中,这部分内存称为 change buffer。change buffer 的大小由系统参数 innodb_change_buffer_max_size 指定,默认为 25,即占整个数据缓冲区的 25%。下次读取磁盘上这个数据页时,读出的数据与 change buffer 内本属于这个数据页的数据合并。关闭 mysqld 服务时,change buffer 的内容会存入 system 表空间的 change buffer。

13.2.2　doublewrite buffer

因为 MySQL 的数据页一般由若干个操作系统数据页构成,在把 MySQL 数据缓冲区中的数据写回至数据文件时,会转换为多个操作系统数据页写入操作。当一个 MySQL 数据页只有部分内容写到磁盘时,若服务器发生故障,会导致数据页内容不一致。为了解决这种问题,MySQL 先把数据缓冲区内容写到磁盘上的 doublewrite buffer,然后再一次写入数据文件。在 MySQL 8.0.20 版本之前,这部分内容存储于 system 表空间。

从 MySQL 8.0.20 版本开始,doublewrite buffer 数据存储于数据目录下的两个文件中:'#ib_16384_0.dblwr'和'#ib_16384_1.dblwr',文件名中的 ib 表示 InnoDB 引擎,16384 表示数据页大小,即 16 KB,0 和 1 表示文件序号。

13.2.3　system 表空间的数据文件

system 表空间可以由一个或多个数据文件构成,默认为一个,名称为 ibdata1。

innodb_data_file_path 系统参数指定 system 表空间的数据文件及属性,其值描述 system 表空间中的每个数据文件的属性,包括名称、大小、自动增长、最大大小。默认名称为 ibdata1,大小为 12 MB,自动增长。自动增长的大小由系统参数 innodb_autoextend_increment 决定,默认为 64 MB。

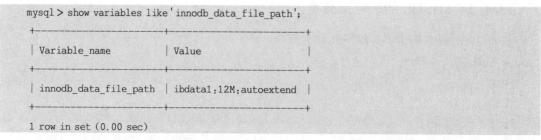

innodb_data_file_path 为静态参数,需要在/etc/my.cnf 中指定新值。

ibdata1 的大小不能修改,若需要扩充 system 表空间大小,可以通过增加数据文件实现。增加数据文件时,先在操作系统中查到当前数据文件的大小,如 ibdata1 的当前大小为 12 MB,

而新加的数据文件为 ibdata2,大小为 20 MB,自动增长,最大大小为 500 MB。在配置文件 /etc/my.cnf 文件的[mysqld]参数组中增加如下一行:

```
[mysqld]
innodb_data_file_path = ibdata1:12M;ibdata2:20M:autoextend:max:500M
```

注意,只有最后一个数据文件才能附加 autoextend 属性。

修改 innodb_data_file_path 的值后,需要重启 mysqld 服务才会生效。

13.3 temporary 表空间

临时表空间(temporary 表空间)存储临时数据,当服务器重启时,临时表空间会删除重建。临时表空间分为两种,会话临时表空间和全局临时表空间。

13.3.1 会话临时表空间

会话临时表空间存放用户创建的临时表,以及执行 SQL 命令时,优化器产生的内部临时表。一个会话最多分配两个会话表空间,一个用于存储用户创建的临时表,一个用于存储优化器针对此用户的 SQL 命令产生的内部临时表。

mysqld 服务启动时,会创建 10 个会话表空间,每个表空间对应的数据文件大小为 5 个数据页(80 KB)。在服务器运行过程中,会根据需要添加新的会话表空间。会话退出连接时,对其分配的会话表空间重新释放给其他会话使用。在整个服务器运行期间,会话表空间的数量不会减少。服务器重启时,会话临时表空间会恢复至 10 个。

可以使用 information_schema.innodb_session_temp_tablespaces 查询会话表空间的信息。

连接数据库后,查询当前会话 id,然后创建临时表 t。

```
mysql> select connection_id();
+-----------------+
| connection_id() |
+-----------------+
|              10 |
+-----------------+
1 row in set (0.00 sec)

mysql> create temporary table t(a int, b char(50));
Query OK, 0 rows affected (0.01 sec)
```

查询此时会话表空间及其使用情况。

```
mysql> select * from information_schema.innodb_session_temp_tablespaces;
+----+------------+------------------------------+-------+----------+-----------+
| ID | SPACE      | PATH                         | SIZE  | STATE    | PURPOSE   |
+----+------------+------------------------------+-------+----------+-----------+
| 10 | 4294501265 | ./#innodb_temp/temp_9.ibt    | 98304 | ACTIVE   | USER      |
| 10 | 4294501266 | ./#innodb_temp/temp_10.ibt   | 81920 | ACTIVE   | INTRINSIC |
|  0 | 4294501257 | ./#innodb_temp/temp_1.ibt    | 81920 | INACTIVE | NONE      |
```

```
|   0 | 4294501258 | ./#innodb_temp/temp_2.ibt | 81920 | INACTIVE | NONE |
|   0 | 4294501259 | ./#innodb_temp/temp_3.ibt | 81920 | INACTIVE | NONE |
|   0 | 4294501260 | ./#innodb_temp/temp_4.ibt | 81920 | INACTIVE | NONE |
|   0 | 4294501261 | ./#innodb_temp/temp_5.ibt | 81920 | INACTIVE | NONE |
|   0 | 4294501262 | ./#innodb_temp/temp_6.ibt | 81920 | INACTIVE | NONE |
|   0 | 4294501263 | ./#innodb_temp/temp_7.ibt | 81920 | INACTIVE | NONE |
|   0 | 4294501264 | ./#innodb_temp/temp_8.ibt | 81920 | INACTIVE | NONE |
+-----+------------+---------------------------+-------+----------+------+
10 rows in set (0.00 sec)
```

可以看到,会话表空间一共有 10 个,以上会话使用了其中两个。最后一列为 USER 的行表示用于存储用户创建的临时表,INTRINSIC 表示用于存储内部临时表。第 1 列为会话的连接 id 标识,第 2 列为表空间 id,第 3、4 列为文件名称和大小(单位为字节),第 5 列表示相应表空间是否正在使用。

临时表空间的数据文件存储在数据目录的子目录'#innodb_temp'中。

```
[root /var/lib/mysql/#innodb_temp 2021-03-02 10:30:37]
# ls
temp_10.ibt  temp_2.ibt  temp_4.ibt  temp_6.ibt  temp_8.ibt
temp_1.ibt   temp_3.ibt  temp_5.ibt  temp_7.ibt  temp_9.ibt
```

13.3.2 全局临时表空间

全局临时表空间存放用户创建的临时表产生的 undo 数据。

全局临时表空间对应的数据文件默认为 ibtmp1。全局临时表空间的数据文件属性由系统参数 innodb_temp_data_file_path 指定。

```
mysql> show variables like 'innodb_temp_data_file_path';
+---------------------------+-----------------------+
| Variable_name             | Value                 |
+---------------------------+-----------------------+
| innodb_temp_data_file_path | ibtmp1:12M:autoextend |
+---------------------------+-----------------------+
1 row in set (0.01 sec)
```

可以看到,这里的表示方法与设置 system 表空间属性的 innodb_data_file_path 参数相似。其初始大小也为 12 MB。

与修改 system 表空间数据文件属性相似,若修改全局临时表空间的数据文件属性,可以在/etc/my.cnf 配置文件中,增加下面一行来设置 innodb_temp_data_file_path 的值。

```
[mysqld]
innodb_temp_data_file_path = ibtmp1:12M;ibtmp2:20m:autoextend:max:500m
```

每次服务器正常关闭时,会删除全局临时表空间的数据文件。服务器重启时,会依据此参数的文件初始大小重建数据文件。

若全局临时表空间在运行过程中,增长得过大,则可以通过重启服务器重建数据文件,使其恢复至初始大小。

13.4 undo 表空间

undo 表空间存储 MySQL 运行过程中产生的 undo 数据（如 delete 操作删除的行，以及 update 操作执行前的原数据），用于 rollback 操作或满足事务一致性和隔离性的要求，undo 数据也称多版本数据。MySQL 5.6 版本开始支持 undo 表空间，在之前的版本中，undo 数据存储于 system 表空间。

13.4.1 默认 undo 表空间

MySQL 8.0 的 undo 数据默认存储于单独的 undo 表空间，undo 表空间的数量最少为 2。MySQL 安装后初始化时，会自动创建两个 undo 表空间，文件名称为 undo_001，undo_002，在数据字典表中记录的表空间名称为 innodb_undo_001，innodb_undo_002，这两个表空间不能删除。undo 数据文件的初始大小决定于数据页大小，对于 16 KB 的默认数据页大小，在 MySQL 8.0.23 版本中，undo 数据文件初始大小为 16 MB。

13.4.2 创建 undo 表空间

从 MySQL 8.0.14 开始，创建 undo 表空间可以使用 create undo tablespace 语句，最多可以增加到 127 个。

增加一个 undo 表空间 undotbs1，其数据文件扩展名要求为.ibu。

```
mysql> create undo tablespace undotbs1 add datafile 'undotbs1.ibu';
```

undo 表空间的数据文件默认存入数据目录，若需指定另外的存储目录，则要把此目录加至 innodb_directories 参数并把目录属主及组改为 mysql:mysql，innodb_directories 参数是只读参数，需要将其加入 my.cnf 文件，设置此参数后需重启服务器。

如下所示，把 innodb_directories 指定为两个目录：/var/lib/mytbs 及/mytbs。

```
[mysqld]
innodb_directories = "/var/lib/mytbs;/mytbs"
```

13.4.3 查询 undo 表空间信息

要得到 undo 表空间的系统信息，可以查询 information_schema.files 系统视图。

```
mysql> select tablespace_name, file_id, file_name,
    -> extent_size * total_extents / 1024 /1024 as "current_size(MB)", status
    -> from information_schema.files
    -> where file_type = 'UNDO LOG';
+-----------------+------------+------------------+------------------+--------+
| TABLESPACE_NAME | FILE_ID    | FILE_NAME        | current_size(MB) | STATUS |
+-----------------+------------+------------------+------------------+--------+
| innodb_undo_001 | 4294967279 | ./undo_001       |     564.00000000 | NORMAL |
| innodb_undo_002 | 4294967278 | ./undo_002       |     656.00000000 | NORMAL |
| undotbs1        | 4294967277 | ./undotbs1.ibu   |      10.00000000 | NORMAL |
+-----------------+------------+------------------+------------------+--------+
3 rows in set (0.00 sec)
```

undo 表空间的数据文件 id 从 4294967279 开始,逐次递减。

13.4.4 截断 undo 表空间

为了避免 undo 表空间无限增长,需要对其执行截断。MySQL 可以使用自动和手动两种方法截断 undo 表空间。

自动截断方式只需要配置系统参数,而 undo 表空间的去活(deactivation)、截断和重新激活等操作都会自动完成。采用自动截断方式,要求至少有两个 undo 表空间,当其中一个设置为去活状态时,还有另一个可用,因为两个默认 undo 表空间都不能删除,所以这个条件会自动满足。

innodb_undo_log_truncate 参数用于设置是否采用自动截断,其默认值即为 on。

innodb_undo_log_truncate 开启后,若表空间大小超过参数 innodb_max_undo_log_size 的设定值(默认为 1 GB),MySQL 会将其标记为截断(truncation),把它的回滚段标记为非激活状态(undo 表空间由多个回滚段构成),不再分配给新的事务,purge 线程释放不再使用的回滚段。所有的回滚段都释放后,把 undo 表空间截断至初始大小,重新激活回滚段至正常可用状态。

若采用手动截断方式,则需要至少 3 个 undo 表空间,以支持可能的自动截断设置。

执行手动截断,先选择一个 undo 表空间,如 undotbs1,执行下面命令将其去活:

```
mysql> alter undo tablespace undotbs1 set inactive;
```

执行以上命令去活后,MySQL 不再把 undotbs1 分配给新的事务,其中的活动事务结束后,purge 线程释放其中的回滚段,截断至其初始大小,将其状态设置为 empty。

要确认其状态为 empty,可以执行下面的命令。

```
mysql> select name, state from information_schema.innodb_tablespaces
    -> where name = 'undotbs1';
+----------+-------+
| name     | state |
+----------+-------+
| undotbs1 | empty |
+----------+-------+
1 row in set (0.00 sec)
```

确认其状态为 empty 后,执行下面的命令将其重新激活。

```
mysql> alter undo tablespace undotbs1 set active;
```

以下几个状态参数,可以查看 undo 表空间的信息。

```
mysql> show status like 'innodb_undo_tablespaces%';
+--------------------------------+-------+
| Variable_name                  | Value |
+--------------------------------+-------+
| Innodb_undo_tablespaces_total    | 3     |
| Innodb_undo_tablespaces_implicit | 2     |
| Innodb_undo_tablespaces_explicit | 1     |
| Innodb_undo_tablespaces_active   | 3     |
+--------------------------------+-------+
4 rows in set (0.00 sec)
```

其中，Innodb_undo_tablespaces_implicit 表示系统初始化时创建的默认 undo 表空间个数，Innodb_undo_tablespaces_explicit 为用户创建的表空间个数，Innodb_undo_tablespaces_active 表示处于激活状态的 undo 表空间个数。

13.4.5　删除 undo 表空间

除了下面过程的最后一步，删除 undo 表空间与对其执行手动截断的步骤相似。

如删除 undotbs1，先对其去活。

```
mysql> alter undo tablespace undotbs1 set inactive;
```

确认其状态为 empty。

```
mysql> select name, state
    -> from information_schema.innodb_tablespaces
    -> where name = 'undotbs1';
+----------+-------+
| name     | state |
+----------+-------+
| undotbs1 | empty |
+----------+-------+
1 row in set (0.00 sec)
```

最后执行删除操作。

```
mysql> drop undo tablespace undotbs1;
```

13.5　file-per-table 表空间

用户在创建表时，可以选择 system、file-per-table 或 general 表空间存储其数据。

system 表空间在 MySQL 系统初始化时创建。file-per-table 表空间的特点是表空间与表是一一对应的，创建表时，即自动产生对应的表空间，删除表时，其对应的表空间和数据文件也会自动删除。而 general 表空间需要用户手动创建，可以存储多个表的数据，要使用 general 表空间，需在创建表时指定表空间的名称。

file-per-table 和 general 表空间都由一个文件构成。

如果不使用 general 表空间，则系统参数 innodb_file_per_table 决定表存储的表空间种类。若为 1 或 on，则存储于 file-per-table 表空间，若为 0 或 off，则存储于 system 表空间，on 为默认值。system 表空间主要用于存储系统数据，用户表不推荐存入 system 表空间。

确认 innodb_file_per_table 的值。

```
mysql> show variables like 'innodb_file_per_table';
+-----------------------+-------+
| Variable_name         | Value |
+-----------------------+-------+
| innodb_file_per_table | ON    |
+-----------------------+-------+
1 row in set (0.00 sec)
```

在 db 数据库创建表 t。

```
mysql> create table db.t(a int, b int);
```

在数据目录的 db 子目录会创建与表同名的文件 t.ibd。

```
[root /var/lib/mysql/db 2021-03-02 15:31:00]
# ls -lh t*
-rw-r-----. 1 mysql mysql 112K Mar  2 15:30 t.ibd
```

查询 information_schema 系统库中的 innodb_tables 和 innodb_tablespaces 表，可以得到相似的结果。

```
mysql> select space, name, space_type
    -> from information_schema.innodb_tables
    -> where name = 'db/t';
+-------+------+------------+
| space | name | space_type |
+-------+------+------------+
|    40 | db/t | Single     |
+-------+------+------------+
1 row in set (0.00 sec)

mysql> select space, name, file_size, space_type
    -> from information_schema.innodb_tablespaces
    -> where name = 'db/t';
+-------+------+-----------+------------+
| space | name | file_size | space_type |
+-------+------+-----------+------------+
|    40 | db/t |    114688 | Single     |
+-------+------+-----------+------------+
1 row in set (0.00 sec)
```

注意这里的表名和表空间名称都是 db/t，即数据库名称附加表名，space 列表示表空间编号，file_size 列表示文件大小（字节），space_type 列表示表空间类型，其可能值为 Single（即 file-per-table 表空间）、System、General、Undo。

修改 innodb_file_per_table 的值为 0，重复以上操作。

```
mysql> drop table db.t;
mysql> set global innodb_file_per_table = 0;
mysql> create table db.t(a int, b int);
```

此时，db 目录下不再创建 t.ibd 文件。

```
[root /var/lib/mysql/db 2021-03-02 15:37:36]
# ls -l t.ibd
ls: cannot access 't.ibd': No such file or directory
```

查询 information_schema 系统数据库中的 innodb_tables 和 innodb_tablespaces 表。

```
mysql> select space, name, space_type from information_schema.innodb_tables
    -> where name = 'db/t';
+-------+------+------------+
```

```
+-------+------+------------+
| space | name | space_type |
+-------+------+------------+
|     0 | db/t | System     |
+-------+------+------------+
1 row in set (0.00 sec)

mysql> select space, name, file_size, space_type
    -> from information_schema.innodb_tablespaces
    -> where name = 'db/t';
Empty set (0.00 sec)
```

从第 1 个查询可以看到，t 表存入 system 表空间。information_schema.innodb_tablespaces 中不保存 system 表空间信息，在这里也不再显示 t 表的相关内容。

13.6　general 表空间

如果使用 file-per-table 表空间，当表的个数太多时，表空间和数据文件也会很多，会造成用于管理文件的资源过多，为了解决此问题，MySQL 5.7.6 引入了 general 表空间。

general 表空间由一个扩展名为 .ibd 的数据文件构成，可以存储多个表的数据，要使用 general 表空间，只需建表时附加 tablespace 子句指定表空间名称。

删除 general 表空间中的表，不会把空间释放回操作系统，只能在表空间内部重用。

创建表空间时，只需指定其名称及数据文件名称。

```
mysql> create tablespace tbs1 add datafile 'tbs1.ibd';
```

在数据目录中创建数据文件。

```
[root /var/lib/mysql 2021-03-02 15:54:30]
# ls -lh tbs*
-rw-r-----. 1 mysql mysql 112K Mar  2 15:54 tbs1.ibd
```

查询新建的表空间信息。

```
mysql> select space, name, file_size, space_type
    -> from information_schema.innodb_tablespaces
    -> where name = 'tbs1';
+-------+------+-----------+------------+
| space | name | file_size | space_type |
+-------+------+-----------+------------+
|     3 | tbs1 |    114688 | General    |
+-------+------+-----------+------------+
1 row in set (0.00 sec)
```

创建 db.t 表，指定存入 tbs1 表空间。

```
mysql> create table db.t(a int, b int) tablespace = tbs1;
```

查询 db.t 的信息。

```
mysql> select space, name, space_type from information_schema.innodb_tables
    -> where name = 'db/t';
+-------+------+------------+
| space | name | space_type |
+-------+------+------------+
|     3 | db/t | General    |
+-------+------+------------+
1 row in set (0.00 sec)
```

查询表所在的表空间。

```
mysql> select it.name, itbs.name, it.space_type
    -> from information_schema.innodb_tables it,
    ->      information_schema.innodb_tablespaces itbs
    -> where it.name = 'db/t'
    -> and itbs.space = it.space;
+------+------+------------+
| name | name | space_type |
+------+------+------------+
| db/t | tbs1 | General    |
+------+------+------------+
1 row in set (0.00 sec)
```

反过来，要查询一个通用表空间包含哪些表，可使用下面命令。

```
mysql> select name from information_schema.innodb_tables
    -> where space =
    -> (
    ->     select space from information_schema.innodb_tablespaces
    ->     where name = 'tbs1'
    -> );
```

为了避免与 file-per-table 表空间混淆，MySQL 不允许把通用表空间的数据文件放置在数据目录下的子目录中。

若要把数据文件放到数据目录以外的目录，需要把相应目录设置为 innodb_directories 系统参数的值，让 MySQL 提前得知这些目录的存在。

如要把数据文件放到/var/lib/mytbs。

创建子目录/var/lib/mytbs，并设置其属主和属组为 mysql:mysql。

```
# mkdir /var/lib/mytbs
# chown mysql:mysql /var/lib/mytbs
```

修改/etc/my.cnf 文件，加入下面一行，使 innodb_directories 包括/var/lib/mytbs。

```
[mysqld]
innodb_directories = "/var/lib/mytbs;/mytbs"
```

重启服务器后，创建表空间 tbs2，使其数据文件存放在/var/lib/mytbs 目录下。

```
mysql> create tablespace tbs2 add datafile '/var/lib/mytbs/tbs2.ibd';
```

13.7 移动表所属的表空间

MySQL 的表可在不同的 general 表空间之间移动,也可在不同种类的表空间之间移动。在表空间之间移动表,使用 alter table 命令。

在 MySQL 的数据字典中,system 和 file_per_table 表空间分别记为 innodb_system 和 innodb_file_per_table。

创建测试表 db.t,默认存储于 innodb_file_per_table 表空间。

```
mysql> create table db.t(a int);
```

将其移至 innodb_system 表空间,即 system 表空间。

```
mysql> alter table db.t tablespace = innodb_system;
```

将其移至 tbs1 表空间,然后再移至 tbs2 表空间,这两个表空间都是 general 表空间。

```
mysql> alter table db.t tablespace = tbs1;
mysql> alter table db.t tablespace = tbs2;
```

第 14 章 B 树索引

B 树结构由波音公司的 Rudolf Bayer 和 Ed McCreight 在 1971 年提出（B 一般认为表示 Balance），广泛应用于数据库管理系统以及文件系统。B 树索引是数据库查询优化最常用、效果最显著的方式。

B 树索引创建在表的列上，会占用存储空间。修改表的索引列数据会同时修改索引中的数据。创建和删除索引不会影响表的数据。查询是否需要使用索引由优化器确定，一般不需要用户参与。

本章主要内容包括：
- 主键索引和普通索引的结构；
- 基于函数的索引；
- 需要创建索引的情形；
- DML 语句对索引的影响；
- 查询索引的系统信息。

14.1 B 树索引能把查询速度提高多少

为了对索引的作用有一个具体感受，我们用单表查询考查使用索引前后的速度差异。

表的数据量足够大，索引的作用才显现出来。下面的 SQL 命令创建存储过程 big_table，执行这个存储过程，会创建一个表 big_table，参数为添加至表中的行数，这个表由 3 个列构成，id1、id2 为整型列，data 为字符串列。其中 id1 列为主键，在 id2 列上创建索引，以测试索引的效果，data 列只是使得数据量快速增长，没有其他作用。

```
delimiter /
CREATE PROCEDURE big_table(cnt int)
BEGIN
    DECLARE i INT DEFAULT 1;
    SET autocommit = 0;
    DROP TABLE IF EXISTS big_table;
    CREATE TABLE big_table(id1 INT PRIMARY KEY, id2 INT, data VARCHAR(30));
    WHILE (i <= cnt) DO
        INSERT INTO big_table VALUES(i, i, CONCAT("record ", i));
        SET i = i + 1;
    END WHILE;
    COMMIT;
    SET autocommit = 1;
END;
/
delimiter ;
```

执行 big_table 存储过程,参数指定为 10 000 000,创建一个 1 000 万行的表。

```
mysql> call big_table(10000000);
Query OK, 0 rows affected (9 min 52.36 sec)
```

然后在 id2 列上创建 B 树索引,命名为 idx_id2。

```
mysql> create index idx_id2 on big_table(id2);
Query OK, 0 rows affected (11.85 sec)
Records: 0  Duplicates: 0  Warnings: 0
```

索引创建完毕后,执行下面的查询,注意提示信息中的执行时间。

```
mysql> select * from big_table where id2 = 100;
+-----+-----+------------+
| id1 | id2 | data       |
+-----+-----+------------+
| 100 | 100 | record 100 |
+-----+-----+------------+
1 row in set (0.01 sec)
```

可以看到,上面查询用时 0.01 s。

把 id2 上的索引禁用后,再次执行同样的查询。

```
mysql> alter table big_table alter index idx_id2 invisible;
Query OK, 0 rows affected (0.01 sec)
Records: 0  Duplicates: 0  Warnings: 0

mysql> select * from big_table where id2 = 100;
+-----+-----+------------+
| id1 | id2 | data       |
+-----+-----+------------+
| 100 | 100 | record 100 |
+-----+-----+------------+
1 row in set (2.16 sec)
```

禁用索引后,查询用时 2.16 s。

此例中,使用索引,查询速度大约可以提高 200 倍。若数据量更大,索引效果会更明显。

下面我们探讨 B 树索引是如何提高查询速度的。

14.2　一个使用索引的例子

在表上添加主键约束时,MySQL 会对主键列自动创建索引,其他普通索引都以主键索引为基础。

英文词典的内容组织形式与主键索引非常相似。为了读者能快速检索单词,词典正文会把所有单词条目按字母排序,大型词典还会在侧面以字母标识出正文条目的起止范围,以进一步提高查询速度。

图 14-1 所示为《韦氏第三版新国际英语大词典》的正文首页及侧面的字母标识。

总结一下,韦氏词典对单词条目的组织形式有以下特点:
- 将单词条目及其释义按照条目字母排序后,依次印刷到各页;

- 把正文各页面按其单词条目的首字母分为13组,每两个字母一组标识于词典侧面。这种组织形式与 MySQL 的主键索引结构非常相似。

词典正文可看作由单词条目及其释义构成的一个两列表。每个正文页面是索引的一个叶节点,侧面是分支节点层,每两个字母标识是一个分支节点。

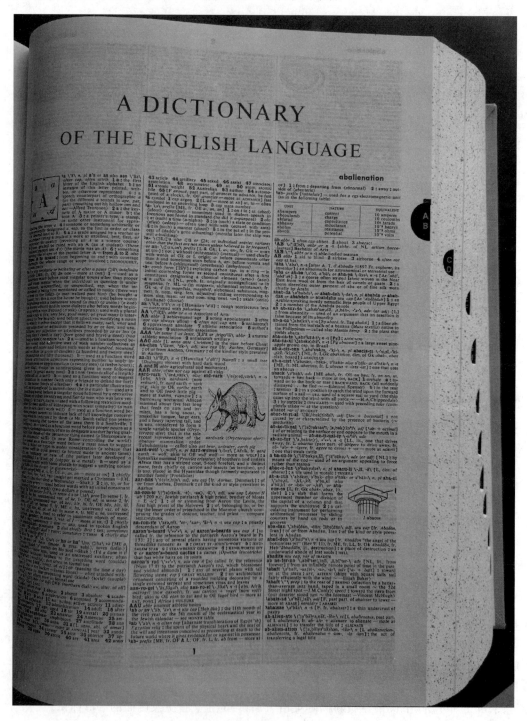

图 14-1　英文词典与主键索引相似

14.3　主键索引的结构

MySQL 会在主键列上自动创建索引,下面用一个具体例子说明主键索引的结构。
创建 employee 表,指定 eno 为主键,然后把 emp 表的 3 列、12 行数据添加至新表。

```
mysql> create table employee
    -> (
    ->      eno int,
    ->      ename varchar(12),
    ->      salary numeric(7, 2),
    ->      primary key(eno)
    -> );
Query OK, 0 rows affected (0.02 sec)

mysql> insert into employee select empno, ename, sal from emp;
Query OK, 12 rows affected (0.01 sec)
Records: 12  Duplicates: 0  Warnings: 0
```

确认 employee 表的数据。

```
mysql> select * from employee;
+------+--------+---------+
| eno  | ename  | salary  |
+------+--------+---------+
| 7369 | SMITH  |  800.00 |
| 7499 | ALLEN  | 1600.00 |
| 7521 | WARD   | 1250.00 |
| 7566 | JONES  | 2975.00 |
| 7654 | MARTIN | 1250.00 |
| 7698 | BLAKE  | 2850.00 |
| 7782 | CLARK  | 2450.00 |
| 7839 | KING   | 5000.00 |
| 7844 | TURNER | 1500.00 |
| 7900 | JAMES  |  950.00 |
| 7902 | FORD   | 3000.00 |
| 7934 | MILLER | 1300.00 |
+------+--------+---------+
12 rows in set (0.00 sec)
```

创建表时,MySQL 会在数据文件中对其分配空间。数据文件被划分为数据页,数据页是 MySQL 读写数据的最小单位,默认为 16 KB,表的行被添加至分到的数据页中。

为了说明方便,我们假定每个数据页只能存放 3 个索引记录(实际当然远多于 3 个)。

以上 12 行记录按照 eno 排序后依次存入分到的 4 个数据页中,这些数据页组成索引的叶节点层,一个叶节点即一个数据页。叶节点层存放了排序后的所有行。

叶节点层的数据页超过 1 个时,产生新的分支节点层,一个分支节点也是一个数据页,存放叶节点层的每个数据页的页号及其中的第 1 个主键值。

分支节点超过 1 个时,又会产生一个新的分支节点层,存放其下一层每个数据页的页号及其中第 1 个键值,这样,最上层总保持一个数据页,这个数据页称为根节点。

上述结构称为 B 树。每层的数据页超过 1 个时,每个数据页也会存储其前后数据页的页号,形成一个双向链表结构,当需要遍历某层时,不需要返回根节点。这种存在双向链表的 B 树结构称为 B+树。各种数据库产品使用的 B 树索引,严格说都是 B+树。

假定叶节点层的数据页号分别为 4、5、6、7,分支节点层的数据页号分别为 8、9,根节点层的数据页号为 10,则 employee 表的主键索引结构如图 14-2 所示。

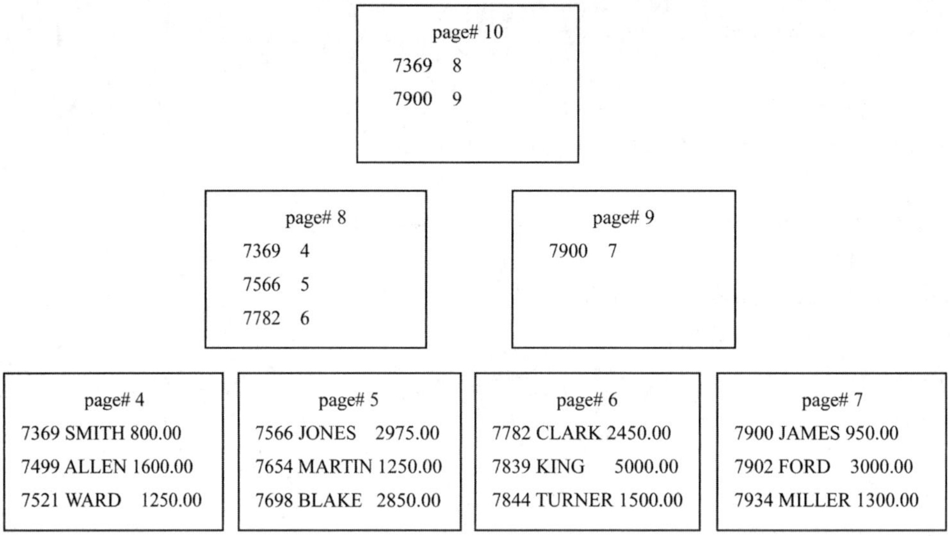

```
                    page# 10
                    7369   8
                    7900   9


        page# 8                       page# 9
        7369   4                      7900   7
        7566   5
        7782   6


  page# 4              page# 5              page# 6              page# 7
  7369 SMITH  800.00   7566 JONES  2975.00  7782 CLARK 2450.00   7900 JAMES  950.00
  7499 ALLEN 1600.00   7654 MARTIN 1250.00  7839 KING  5000.00   7902 FORD  3000.00
  7521 WARD  1250.00   7698 BLAKE  2850.00  7844 TURNER 1500.00  7934 MILLER 1300.00
```

图 14-2 主键索引结构

假定执行查询 select * from emp where empno = 7839,如何利用以上主键索引呢?

① 读取索引根节点,即 10 号数据页,读出其中的两行数据。

② 条件为 empno = 7839,9 号数据页中最小键值为 7900,判断出下一层应读取 8 号数据页。

③ 读取 8 号数据页,判断出下一层应读取 6 号数据页。

④ 读取 6 号数据页,其中的第 2 行即满足条件。

利用主键索引,每层需要读取一次,完成以上查询操作,一共需要读取 3 次。由根节点遍历至任意一个叶节点的读取次数都相同。

主键索引的名称总为 PRIMARY。如果未对表添加主键约束,MySQL 会把存在非空唯一约束的第 1 个列作为主键对待,索引名称与约束名称相同,若对约束未命名,则索引名称为列名。若表上既没有主键约束也没有非空唯一约束,MySQL 会添加一个新列 DB_ROW_ID 作为主键列,以类似自增列的处理方式自动对其添加值,这个索引名称也为 PRIMARY。

查看表上的索引情况,可以执行 show index from *table_name* 命令。如下面的命令可查看 employee 表上的主键索引,其中的注释为作者所加。

```
mysql> show index from employee \G
*************************** 1. row ***************************
        Table: employee        # 索引列值是否重复,1为重复,0为唯一
   Non_unique: 0               
     Key_name: PRIMARY          # 索引名称,主键索引名称总为 PRIMARY
 Seq_in_index: 1                # 列在复合索引中的序号
  Column_name: eno              # 索引列名称
    Collation: A                # 索引是升序还是降序,A 为升序,D 为降序
  Cardinality: 12               # 索引列值中不重复的个数
     Sub_part: NULL             # 字符串列被索引的字符数(若索引字符串列一部分)
       Packed: NULL             # 索引列值是否被压缩
         Null:                  # 索引列值是否包含 NULL 值,YES 表示包含,空表示不包含
   Index_type: BTREE            # 索引类型,对于 InnoDB 引擎,总是 BTREE
      Comment:                  # 索引属性(如 disabled),对于 InnoDB 引擎,总是空
Index_comment:                  # 创建索引时,用 comment 属性附加的注释
      Visible: YES              # 索引是否可见
   Expression: NULL             # 基于函数索引的函数表达式
1 row in set (0.00 sec)
```

附加 extended 选项,执行 show index from *table_name*,可以显示索引的所有列。对于主键索引,可以确认其包含了表的所有列。另外还包含了系统对表附加的两个列:DB_TRX_ID 列存储正在修改此行的事务 ID;当修改一个行时,对应的旧版本数据会存入 undo 表空间,DB_ROLL_PTR 列存储指向这个旧版本数据的指针。

下面附加 extended 选项,执行 show index 命令。

```
mysql> show extended index from employee \G
*************************** 1. row ***************************
        Table: employee
   Non_unique: 0
     Key_name: PRIMARY
 Seq_in_index: 1
  Column_name: eno
    Collation: A
  Cardinality: 12
...
*************************** 2. row ***************************
        Table: employee
   Non_unique: 0
     Key_name: PRIMARY
 Seq_in_index: 2
  Column_name: DB_TRX_ID
    Collation: A
  Cardinality: NULL
...
```

```
*************************** 3. row ***************************
        Table: employee
   Non_unique: 0
     Key_name: PRIMARY
 Seq_in_index: 3
  Column_name:DB_ROLL_PTR
...
*************************** 4. row ***************************
        Table: employee
   Non_unique: 0
     Key_name: PRIMARY
 Seq_in_index: 4
  Column_name: ename
...
*************************** 5. row ***************************
        Table: employee
   Non_unique: 0
     Key_name: PRIMARY
 Seq_in_index: 5
  Column_name: salary
...
5 rows in set (0.01 sec)
```

其他数据库产品,表和索引的数据一般是分开存放的。表中的行依其添加的时间先后存储,并无排序。这种行无序存储的表一般称为堆表(heap table)。若表的行指定为以索引方式排序后存储,Oracle 称为索引组织表,SQL Server 称为聚集索引(clustered index)。

通过上述分析,会发现,MySQL 的行总是按照主键索引组织存储的,MySQL 中并不存在堆表这种结构(限于默认的 InnoDB 引擎),MySQL 也把这种索引称为聚集索引。

14.4　普通索引的结构

普通索引的叶节点存储两项内容:排序后的索引列值及其所在记录的主键值。

普通索引和主键索引的分支节点存储内容是相似的,区别在于叶节点。主键索引的叶节点存储了整行,而普通索引只存储了索引列和主键列的值。

下面以实例描述普通索引的结构。

执行下面的命令,在 ename 列上创建索引。

```
mysql> create index idx_ename on employee(ename);
```

叶节点层存储下面按 ename 列排序的 2 列 12 行数据(ename 和 eno 列)。

```
mysql> select ename, eno from employee order by ename;
+--------+------+
| ename  | eno  |
+--------+------+
| ALLEN  | 7499 |
```

```
| BLAKE  | 7698 |
| CLARK  | 7782 |
| FORD   | 7902 |
| JAMES  | 7900 |
| JONES  | 7566 |
| KING   | 7839 |
| MARTIN | 7654 |
| MILLER | 7934 |
| SMITH  | 7369 |
| TURNER | 7844 |
| WARD   | 7521 |
+--------+------+
12 rows in set (0.00 sec)
```

为了说明方便,假定每个数据页只能存放 3 个索引记录。12 行记录按照 ename 排序后依次存入分到的 4 个数据页,这些数据页组成索引的叶节点层,假定其页号为 11、12、13、14,分支节点页号为 15、16,根节点页号为 17。

idx_ename 索引的各层数据页内容,如图 14-3 所示。

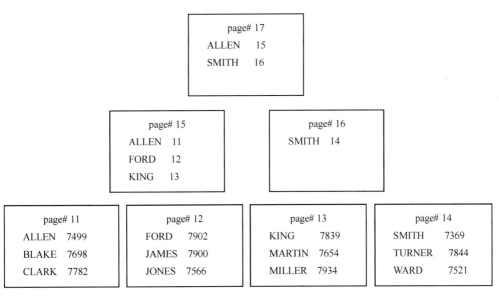

图 14-3 普通索引结构

执行 show extended index 命令,显示 idx_ename 包含的所有列,可以看到其包含了索引列 ename 和主键列 eno。

```
mysql> show extended index from employee where key_name = 'idx_ename'\G
*************************** 1. row ***************************
        Table: employee
   Non_unique: 1
     Key_name: idx_ename
 Seq_in_index: 1
```

```
      Column_name: ename
         Collation: A
       Cardinality: 12
...
*************************** 2. row ***************************
        Table: employee
   Non_unique: 1
     Key_name: idx_ename
 Seq_in_index: 2
  Column_name: eno
...
2 rows in set (0.00 sec)
```

假定执行 select * from emp where ename = 'MARTIN'，如何利用以上索引完成查询呢？

① 首先读取索引根节点，即 17 号数据页，读出其中的两行数据。
② 16 号数据页的最小键值为 SMITH，判断出下一层应读取 15 号数据页。
③ 读取 15 号数据页，可判断出下一层应读取 13 号数据页。
④ 读取 13 号数据页，其中的第 2 行即满足条件，读出其主键值 7654。
⑤ 由主键值 7654 读取主键索引。
⑥ 读取主键索引的根节点，即 10 号数据页，判断出下一层应读取 8 号数据页。
⑦ 读取 8 号数据页，可判断出下一层应读取 5 号数据页。
⑧ 读取 5 号数据页，其第 2 行即满足条件。
⑨ 读出第 2 行整行 7654 MARTIN 1250.00，查询结束。

两个索引，每一层都要读取一次，读出数据，共需读取 6 次。

如果表不存在主键约束，也没有其他非空唯一约束，则 MySQL 会自动添加一个新列 DB_ROW_ID 作为主键列，普通索引的叶节点由索引列与 DB_ROW_ID 列构成。

删除 employee 表上的主键约束，然后验证以上结论。

```
mysql> alter table employee drop constraint `primary`;
Query OK, 12 rows affected (0.02 sec)
Records: 12  Duplicates: 0  Warnings: 0

mysql> show extended index from employee\G
*************************** 1. row ***************************
        Table: employee
   Non_unique: 0
     Key_name: PRIMARY
 Seq_in_index: 1
  Column_name: DB_ROW_ID
    Collation: A
  Cardinality: NULL
...
```

```
*************************** 2. row ***************************
        Table: employee
   Non_unique: 0
     Key_name: PRIMARY
 Seq_in_index: 2
  Column_name: DB_TRX_ID
...
*************************** 3. row ***************************
        Table: employee
   Non_unique: 0
     Key_name: PRIMARY
 Seq_in_index: 3
  Column_name: DB_ROLL_PTR
...
*************************** 4. row ***************************
        Table: employee
   Non_unique: 0
     Key_name: PRIMARY
 Seq_in_index: 4
  Column_name: eno
...
*************************** 5. row ***************************
        Table: employee
   Non_unique: 0
     Key_name: PRIMARY
 Seq_in_index: 5
  Column_name: ename
...
*************************** 6. row ***************************
        Table: employee
   Non_unique: 0
     Key_name: PRIMARY
 Seq_in_index: 6
  Column_name: salary
...
*************************** 7. row ***************************
        Table: employee
   Non_unique: 1
     Key_name: idx_ename
 Seq_in_index: 1
  Column_name: ename
    Collation: A
  Cardinality: 12
...
```

```
*************************** 8. row ***************************
        Table: employee
   Non_unique: 1
     Key_name: idx_ename
 Seq_in_index: 2
  Column_name: DB_ROW_ID
    Collation: A
...
8 rows in set (0.00 sec)
```

删除主键约束后，employee 表的统计信息也失效了，以上结果的 Cardinality 列值均为 NULL，可以执行 analyze table 重建统计信息，具体用法请参考第 16 章统计信息相关内容。

```
mysql > analyze table employee;
```

14.5 索引能够提高查询速度的原因

如果不考虑选择执行计划的时间，查询时间主要由读取磁盘时间、读取内存时间和 CPU 处理时间决定，而其中最主要的部分是读取磁盘时间，读取磁盘时间由读取次数决定，每次读取一个数据页。

如果没有索引，读取数据时，需要执行全表扫描，如果一个表的行需要 1 000 个数据页存储，则全表扫描需要读取 1 000 次。

在使用索引的情况下，如果索引键值重复率不高，则读取磁盘次数主要由索引层数决定，使用主键作为查询条件，则只需要使用主键索引，使用普通索引列作为查询条件，需要先读取普通索引，再读取主键索引。

索引一般不超过 3 层，即使使用普通和主键两个索引，读取次数一般也不超过 6 次。

14.6 需要创建索引的情况

对于大数据量（记录行数一般大于 100 万）的单表查询，频繁用作查询条件的列，若列值重复率低，一般需要对其创建索引。

多表连接查询中，一般对在大表中用作连接条件的列创建索引。

14.7 如何知道一个查询是否使用了索引

执行 SQL 命令之前，MySQL 根据数据分布情况制订执行计划。可以通过查看执行计划确认一个查询是否使用了索引。查看执行计划，只需在查询命令之前附加 explain 关键字。如果显示结果中的 type 列值为 ALL，则表示未使用索引，若为其他值，则一般表示使用了索引，其 key 列值为索引的名称。

关闭 id2 列上的索引后，查看 big_table 的执行计划。

```
mysql> alter table big_table alter index idx_id2 invisible;
Query OK, 0 rows affected (0.02 sec)
Records: 0  Duplicates: 0  Warnings: 0

mysql> explain select * from big_table where id2 = 100\G
*************************** 1. row ***************************
           id: 1
  select_type: SIMPLE
        table: big_table
   partitions: NULL
         type: ALL
possible_keys: NULL
          key: NULL
      key_len: NULL
          ref: NULL
         rows: 9761487
     filtered: 0.00
        Extra: Using where
1 row in set, 1 warning (0.00 sec)
```

由上面显示结果中的 type 列为 ALL 可知,查询计划使用了全表扫描,未使用索引。

开启索引 idx_id2 后,再次查看执行计划。

```
mysql> alter table big_table alter index idx_id2 visible;
Query OK, 0 rows affected (0.02 sec)
Records: 0  Duplicates: 0  Warnings: 0

mysql> explain select * from big_table where id2 = 100\G
*************************** 1. row ***************************
           id: 1
  select_type: SIMPLE
        table: big_table
   partitions: NULL
         type: ref
possible_keys: idx_id2
          key: idx_id2
      key_len: 5
          ref: const
         rows: 1
     filtered: 100.00
        Extra: NULL
1 row in set, 1 warning (0.00 sec)
```

可以发现,此时使用了索引 idx_id2。

14.8　不使用索引的情况

对于下面几种情况，MySQL 不会使用索引。
- 字符串模糊查询时，通配符在第一个位置。
- 查询条件对索引列值使用了函数。这种情况下，可以创建基于函数的索引。

若查询条件涉及的索引列值重复率高，可能使用全表扫描效率反而更高。对这种情况，MySQL 尚缺乏判断，一般只要有索引，MySQL 就会使用。

14.9　DML 语句对索引的影响

若表存在索引，对表执行 DML 操作，即执行 insert、delete 与 update 命令时，也会对索引进行同步修改，对索引的修改也会产生重做数据和 undo 数据。下面分别说明这 3 种语句对索引的影响。

14.9.1　insert 语句对索引的影响

为了保持索引列值的顺序，添加索引记录的数据页及数据页中的位置是确定的。

若新的索引列值添加到的叶节点数据页已经没有空闲空间，则会把此数据页内的数据均分为二，一份留到原数据页，一份移至另外一个空数据页中，同时把数据页中指向前后数据页的指针做相应修改。

14.9.2　delete 语句对索引的影响

对表的记录执行 delete 操作时，只是将其标记为删除，其占用数据页的空间并未释放，当有新行加入此数据页时，这些标记为删除的空间才会释放，从而这些空间可以被重用。与此相似，被删除的记录在索引中对应的数据也会在数据页中标记为删除，当有新的索引记录加入此数据页时，这些标记为删除的索引行占用的空间才会释放。

14.9.3　update 语句对索引的影响

对索引列值执行 update 操作，相当于在索引中对 update 操作之前的原行先执行 delete 操作，然后再把 update 之后的新行添加(insert)至索引。

14.10　基于函数的索引

若查询条件需要对列进行函数作用，则此列上的普通 B 树索引不会被用到。

在 emp 表的 hiredate 列上创建索引。

```
mysql> create index idx_hiredate on emp(hiredate);
```

若以 month(hiredate) 取出其月份作为查询条件，通过查看其执行计划，可以发现 type 列为 ALL，即使用了全表扫描，并未使用 hiredate 上的索引。

```
mysql> explain select * from emp where month(hiredate) = 12 \G
*************************** 1. row ***************************
           id: 1
  select_type: SIMPLE
        table: emp
   partitions: NULL
         type: ALL
...
         rows: 12
     filtered: 100.00
        Extra: Using where
1 row in set, 1 warning (0.01 sec)
```

应对这种情况,可创建基于函数的索引。基于函数的索引是 MySQL 8.0.13 引入的功能。

在 hiredate 列上创建基于 month() 函数的索引,重新查看执行计划,此时使用了索引。

```
mysql> create index idx_month on emp((month(hiredate)));
Query OK, 0 rows affected (0.03 sec)
Records: 0  Duplicates: 0  Warnings: 0

mysql> explain select * from emp where month(hiredate) = 12 \G
*************************** 1. row ***************************
           id: 1
  select_type: SIMPLE
        table: emp
   partitions: NULL
         type: ref
possible_keys: idx_month
          key: idx_month
      key_len: 5
          ref: const
         rows: 3
     filtered: 100.00
        Extra: NULL
1 row in set, 1 warning (0.00 sec)
```

14.11　设置索引的可见性

可以指定索引对优化器是否可见(visible/invisible)。设置为 invisible 后,对优化器不可见,但对表执行 DML 操作时,MySQL 仍然会同步更新索引中的数据。索引的这种属性使测试索引对查询效率的影响时,可以避免将其反复创建、删除。

切换索引 idx_ename 为不可见、可见,可以执行下面的命令。

```
mysql> alter table emp alter index idx_ename invisible;
mysql> alter table emp alter index idx_ename visible;
```

14.12 多列索引

多列索引可以把某些查询限制于索引，而不访问表，特别是在涉及排序的场合，可以顺序访问索引记录，避免排序。

在 emp 表的 ename 和 sal 列创建复合索引 idx_ename_sal。

```
mysql> create index idx_ename_sal on emp(ename, sal);
```

若查询操作只涉及 ename 和 sal 列，直接访问索引即可，如下面的执行计划所示。

```
mysql> explain select ename, sal from emp order by ename\G
*************************** 1. row ***************************
           id: 1
  select_type: SIMPLE
        table: emp
   partitions: NULL
         type: index
possible_keys: NULL
          key: idx_ename_sal
      key_len: 48
          ref: NULL
         rows: 12
     filtered: 100.00
        Extra: Using index
1 row in set, 1 warning (0.00 sec)
```

type 列为 index、Extra 列为 Using index，这两列都表示只访问了索引，key 列标识索引名称。

对于多列索引，若查询的列没有被索引全部包含，使用索引首列作为查询条件，此索引的作用与单列索引相同，若使用非首列作为查询条件，则不会使用多列索引。

下面两个查询的 job 列不在索引中，第 1 个查询使用 ename 作为查询条件，可以看到使用了索引，第 2 个查询使用了 sal 作为查询条件，索引未使用。

```
mysql> explain select ename, job, sal from emp where ename = 'SMITH'\G
*************************** 1. row ***************************
           id: 1
  select_type: SIMPLE
        table: emp
   partitions: NULL
         type: ref
possible_keys: idx_ename_sal
          key: idx_ename_sal
      key_len: 43
          ref: const
         rows: 1
     filtered: 100.00
```

```
                  Extra: NULL
1 row in set, 1 warning (0.00 sec)

mysql> explain select ename, job, sal from emp where sal = 800\G
*************************** 1. row ***************************
           id: 1
  select_type: SIMPLE
        table: emp
   partitions: NULL
         type: ALL
...
         rows: 12
     filtered: 10.00
        Extra: Using where
1 row in set, 1 warning (0.00 sec)
```

14.13 约束与索引

创建主键、唯一以及外键约束时，MySQL 会自动对约束所在的列创建索引。

主键约束名称总是 PRIMARY，其索引名称也总是 PRIMARY。若附加唯一或外键约束时指定了名称，则索引名称与约束名称相同，若未指定名称，则索引名称与列名相同。

14.14 查询索引的系统信息

要得到索引的系统信息，有下面几种方法：
- 执行 show create table 命令，得到索引名称及其所在列。
- 执行 show index from 命令，得到索引详细信息。
- 查询 information_schema.innodb_indexes，得到索引的存储相关信息。

14.14.1 show create table

如果只是查看一个表的哪个列存在索引，可以执行 show create table *tbl_name*，如查看 dept 表上的索引情况。

```
mysql> show create table dept \G
*************************** 1. row ***************************
       Table: dept
Create Table: CREATE TABLE `dept` (
  `deptno` decimal(2,0) NOT NULL,
  `dname` varchar(14) DEFAULT NULL,
  `loc` varchar(13) DEFAULT NULL,
  PRIMARY KEY (`deptno`),
  KEY `idx_dname` (`dname`)
) ENGINE = InnoDB DEFAULT CHARSET = utf8mb4 COLLATE = utf8mb4_0900_ai_ci
1 row in set (0.00 sec)
```

查询结果中的 PRIMARY KEY,一方面标识了主键约束,另一方面也指出了主键列上自动创建的索引。查询结果中的 KEY 关键字与 INDEX 同义,标识了除主键索引外的普通索引名称及所在列名。

14.14.2 show index from

若查看索引更详细的信息,可以执行 show index from *tbl_name*。

```
mysql> show index from dept \G
*************************** 1. row ***************************
        Table: dept
   Non_unique: 0
     Key_name: PRIMARY
 Seq_in_index: 1
  Column_name: deptno
    Collation: A
  Cardinality: 4
     Sub_part: NULL
       Packed: NULL
         Null:
   Index_type: BTREE
      Comment:
Index_comment:
      Visible: YES
   Expression: NULL
1 rows in set (0.00 sec)
```

14.14.3 查询 information_schema.innodb_indexes

innodb_indexes 主要用来查看索引的存储空间信息,包含以下列。

- index_id:索引编号。
- name:索引名称。
- table_id:索引所在表的编号,可以由 innodb_tables 查到表的其他信息。
- type:索引类型。0 为非唯一普通索引,1 为自动附加主键上的索引,2 为唯一普通索引,3 为主键索引,32 为全文索引,64 为空间索引,128 为虚拟列或基于函数的索引。
- n_fields:索引包含的列数。
- page_no:索引根节点所在的数据页页号。
- space:索引所在的表空间编号,可由 innodb_tablespaces_brief 得到完整的文件名称。
- merge_threshold:索引数据页的数据所占空间低于此值的时候,尝试与其相邻数据页合并,默认为 50%。

结合 innodb_indexes、innodb_tables、innodb_tablespaces_brief 3 个视图,查询 db 数据库中 emp 表上的 idx_ename 索引信息。

```
mysql> select i.name index_name, i.n_fields columns,
    ->        t.name table_name, s.path file_name, i.page_no
    -> from
    -> information_schema.innodb_indexes i,
    -> information_schema.innodb_tables t,
    -> information_schema.innodb_tablespaces_brief s
    -> where i.name = 'idx_ename' and t.name = 'db/emp' and
    -> i.table_id = t.table_id and
    -> i.space = s.space;
+------------+---------+------------+--------------+---------+
| index_name | columns | table_name | file_name    | page_no |
+------------+---------+------------+--------------+---------+
| idx_ename  |       2 | db/emp     | ./db/emp.ibd |       6 |
+------------+---------+------------+--------------+---------+
1 row in set (0.01 sec)
```

第 15 章 执 行 计 划

执行计划是 MySQL 执行 SQL 语句时,依据软硬件环境及数据状态制定的执行步骤。能够准确解读执行计划是进行 SQL 优化的基础。

本章主要内容包括:
- 如何得到执行计划;
- 如何解读执行计划。

15.1 执行计划简单示例

MySQL 使用 explain 命令得到查询语句的执行计划。

下面是一个简单示例。

```
mysql> explain select * from emp where ename = 'SMITH'\G
*************************** 1. row ***************************
           id: 1
  select_type: SIMPLE
        table: emp
   partitions: NULL
         type: ALL
possible_keys: NULL
          key: NULL
      key_len: NULL
          ref: NULL
         rows: 12
     filtered: 10.00
        Extra: Using where
1 row in set, 1 warning (0.01 sec)
```

下面解释执行计划中各列的意义。
- id:查询编号。
- select_type:查询类型。
- table:操作的表名。
- partitions:对于分区表,指扫描的分区名称,对于普通表则为 null。
- type:读取表数据时的类型。
- possible_keys:可用的索引。
- key:实际使用的索引。
- key_len:使用索引时,需要读取的索引行的长度。对于多列索引,用于确定实际使用的列数。对于存储空间固定的数据类型,此值也固定,如 int 总是占 4 字节,此值也总为 4。对于字符串类型,此值为最大可能大小,如使用字符集 utf8mb4,每个字符最多需要使用 4 字节,若列类型为 varchar(10),则所需最大空间为 40 字节(4 字节×10),对于变长字符串类型,还需加上用于存储实际长度的 2 字节空间。对于可以为 NULL 的列,需要再加上 1 字节用于表示列值是否确实为 NULL。总结一下,若 int 列未附加 NOT NULL 约束,则 key_len 为 5,若 varchar(10) 列未附加 NOT NULL 约束,使用字符集 utf8mb4 时的 key_len 为 43。

- ref:使用索引时,在 where 条件的比较表达式中使用的列名或常量(比较运算符右侧)。若此值为 func,则表示使用了函数表达式的返回值,可执行 show warnings 得到函数表达式。
- rows:要读取的行数估值。
- filtered:满足条件的行数与读取的总行数之间的百分比,最高为 100,此值越大越好。
- Extra:关于执行计划的额外信息。

下面对以上几个主要列进行说明。

15.2 执行计划的 select_type 属性

对于非查询语句,select_type 为其语句类别,如 DELETE 或 UPDATE。对于查询语句,select_type 的可能值主要包括以下几个。

(1) SIMPLE

表示查询未使用子查询、并等操作,单表查询和普通的连接查询一般为这种形式。

下面是一个简单的单表查询。

```
mysql> explain select * from emp where ename = 'SMITH'\G
*************************** 1. row ***************************
           id: 1
  select_type: SIMPLE
        table: emp
   partitions: NULL
         type: ALL
...
        Extra: Using where
1 row in set, 1 warning (0.00 sec)
```

下面示例是两个表的连接查询,两个表的 select_type 均为 simple。

```
mysql> explain select e.ename, d.dname from emp e, dept d where e.deptno = d.deptno
    -> \G
*************************** 1. row ***************************
           id: 1
  select_type: SIMPLE
        table: d
   partitions: NULL
         type: ALL
possible_keys: PRIMARY
          key: NULL
```

```
            key_len: NULL
                ref: NULL
               rows: 4
           filtered: 100.00
              Extra: NULL
*************************** 2. row ***************************
                 id: 1
        select_type: SIMPLE
              table: e
         partitions: NULL
               type: ref
      possible_keys: fk_deptno
                key: fk_deptno
            key_len: 2
                ref: db.d.deptno
               rows: 4
           filtered: 100.00
              Extra: NULL
2 rows in set, 1 warning (0.00 sec)
```

(2) PRIMARY

使用子查询时,PRIMARY 指外层的主查询,如下面相关子查询示例所示。

```
mysql> explain select ename, (select dname from dept where deptno = emp.deptno)
    -> from emp \G
*************************** 1. row ***************************
                 id: 1
        select_type: PRIMARY
              table: emp
         partitions: NULL
               type: ALL
...
               rows: 12
           filtered: 100.00
              Extra: NULL
*************************** 2. row ***************************
                 id: 2
        select_type: DEPENDENT SUBQUERY
              table: dept
         partitions: NULL
               type: eq_ref
      possible_keys: PRIMARY
                key: PRIMARY
            key_len: 1
                ref: db.emp.deptno
```

```
            rows: 1
        filtered: 100.00
           Extra: NULL
2 rows in set, 2 warnings (0.00 sec)
```

(3) DEPENDENT SUBQUERY

表示子查询,且类型为相关子查询,示例如上。

(4) SUBQUERY

表示子查询,且类型为非相关子查询。下面示例在 select 子句中使用非相关子查询。

```
mysql> explain select ename,(select max(sal) from emp) from emp\G
*************************** 1. row ***************************
              id: 1
     select_type: PRIMARY
           table: emp
      partitions: NULL
            type: ALL
...
            rows: 12
        filtered: 100.00
           Extra: NULL
*************************** 2. row ***************************
              id: 2
     select_type: SUBQUERY
           table: emp
      partitions: NULL
            type: ALL
...
            rows: 12
        filtered: 100.00
           Extra: NULL
2 rows in set, 1 warning (0.00 sec)
```

在 where 子句中使用非相关子查询。

```
mysql> explain select * from emp where sal = (select max(sal) from emp)\G
*************************** 1. row ***************************
              id: 1
     select_type: PRIMARY
           table: emp
      partitions: NULL
            type: ALL
...
            rows: 12
        filtered: 10.00
           Extra: Using where
```

```
*************************** 2. row ***************************
           id: 2
  select_type: SUBQUERY
        table: emp
     partitions: NULL
         type: ALL
...
         rows: 12
     filtered: 100.00
        Extra: NULL
2 rows in set, 1 warning (0.01 sec)
```

(5) DERIVED

from 子句中的子查询。这种情况下，主查询执行计划的 table 列值为 < derived$n$ >，n 为子查询步骤的 id 号。

```
mysql> explain select * from
    -> (select deptno, sum(sal) from emp group by deptno) t \G
*************************** 1. row ***************************
           id: 1
  select_type: PRIMARY
        table: < derived2 >
   partitions: NULL
         type: ALL
...
         rows: 12
     filtered: 100.00
        Extra: NULL
*************************** 2. row ***************************
           id: 2
  select_type: DERIVED
        table: emp
   partitions: NULL
         type: index
possible_keys: fk_deptno
          key: fk_deptno
      key_len: 2
          ref: NULL
         rows: 12
     filtered: 100.00
        Extra: NULL
2 rows in set, 1 warning (0.00 sec)
```

(6) UNION, UNION RESULT

执行并操作时，两个查询的 select_type 分别为 PRIMARY 和 UNION。最后结果的 select_type 为 UNION RESULT，table 列值为 < union1,2 >，1 和 2 为两个查询的 id 值。

```
mysql> explain select * from emp where deptno = 10
    -> union
    -> select * from emp where deptno = 20 \G
*************************** 1. row ***************************
           id: 1
  select_type: PRIMARY
        table: emp
    partitions: NULL
         type: ref
possible_keys: fk_deptno
          key: fk_deptno
      key_len: 2
          ref: const
         rows: 3
     filtered: 100.00
        Extra: NULL
*************************** 2. row ***************************
           id: 2
  select_type: UNION
        table: emp
    partitions: NULL
         type: ref
possible_keys: fk_deptno
          key: fk_deptno
      key_len: 2
          ref: const
         rows: 3
     filtered: 100.00
        Extra: NULL
*************************** 3. row ***************************
           id: NULL
  select_type: UNION RESULT
        table: <union1,2>
    partitions: NULL
         type: ALL
...
         rows: NULL
     filtered: NULL
        Extra: Using temporary
3 rows in set, 1 warning (0.00 sec)
```

15.3 执行计划的 type 属性

type 列指表的访问方式，或表连接的方式。

（1）ALL

访问表时使用了全表扫描。当查询条件中的列上没有可用索引时，会使用这种方式。ALL 是效率最低、应尽量避免的访问方式，ALL 专门使用大写以引起注意。

```
mysql> explain select * from emp where ename = 'SMITH'\G
*************************** 1. row ***************************
           id: 1
  select_type: SIMPLE
        table: emp
   partitions: NULL
         type: ALL
...
         rows: 12
     filtered: 10.00
        Extra: Using where
1 row in set, 1 warning (0.00 sec)
```

（2）const

查询条件使用了主键列（或列上存在唯一约束），与常量进行等值比较。

```
mysql> explain select * from emp where empno = 7369\G
*************************** 1. row ***************************
           id: 1
  select_type: SIMPLE
        table: emp
   partitions: NULL
         type: const
possible_keys: PRIMARY
          key: PRIMARY
      key_len: 4
          ref: const
         rows: 1
     filtered: 100.00
        Extra: NULL
1 row in set, 1 warning (0.00 sec)
```

（3）index

查询操作只对索引执行了全扫描，一般是索引包含了需要读取的所有列。此时，Extra 列的值为 Using index，也表示只扫描了索引。

对 emp 表的 ename、sal 列创建复合索引 idx_ename_sal。

```
mysql> create index idx_ename_sal on emp(ename, sal);
```

执行下面的查询，需要读取的列都包含在 idx_ename_sal 索引中。

```
mysql> explain select ename, sal from emp\G
*************************** 1. row ***************************
           id: 1
  select_type: SIMPLE
        table: emp
   partitions: NULL
         type: index
possible_keys: NULL
          key: idx_ename_sal
      key_len: 48
          ref: NULL
         rows: 12
     filtered: 100.00
        Extra: Using index
1 row in set, 1 warning (0.00 sec)
```

(4) ref

查询操作使用了非唯一普通索引。

```
mysql> explain select ename, sal, job from emp where ename = 'SMITH'\G
*************************** 1. row ***************************
           id: 1
  select_type: SIMPLE
        table: emp
   partitions: NULL
         type: ref
possible_keys: idx_ename_sal
          key: idx_ename_sal
      key_len: 43
          ref: const
         rows: 1
     filtered: 100.00
        Extra: NULL
1 row in set, 1 warning (0.00 sec)
```

(5) eq_ref

在表连接查询中,连接条件列是第 2 个被访问表的主键列(或附加非空唯一约束的列),每次访问第 2 个表时,均使用主键索引返回最多一行。

```
mysql> explain select e.ename, m.ename from emp e, emp m where e.mgr = m.empno \G
*************************** 1. row ***************************
           id: 1
  select_type: SIMPLE
        table: e
   partitions: NULL
         type: ALL
...
```

```
            rows: 12
        filtered: 100.00
           Extra: Using where
*************************** 2. row ***************************
              id: 1
     select_type: SIMPLE
           table: m
      partitions: NULL
            type: eq_ref
   possible_keys: PRIMARY
             key: PRIMARY
         key_len: 2
             ref: db.e.mgr
            rows: 1
        filtered: 100.00
           Extra: NULL
2 rows in set, 1 warning (0.00 sec)
```

（6）range

查询条件包含索引列,且会返回多个值,如使用 or、in、between … and 等比较方式。下面的示例在条件中包含了比较运算符">"。

```
mysql> explain select * from emp where empno > 7500\G
*************************** 1. row ***************************
              id: 1
     select_type: SIMPLE
           table: emp
      partitions: NULL
            type: range
   possible_keys: PRIMARY
             key: PRIMARY
         key_len: 2
             ref: NULL
            rows: 10
        filtered: 100.00
           Extra: Using where
1 row in set, 1 warning (0.01 sec)
```

以 in 关键字作为条件。

```
mysql> explain select * from emp where empno in(7369,7698)\G
*************************** 1. row ***************************
              id: 1
     select_type: SIMPLE
           table: emp
      partitions: NULL
            type: range
```

```
    possible_keys: PRIMARY
              key: PRIMARY
          key_len: 2
              ref: NULL
             rows: 2
         filtered: 100.00
            Extra: Using where
1 row in set, 1 warning (0.00 sec)
```

(7) system

查询的表只有一个常量时,使用这种方式。

```
mysql> explain select * from (select 1) a\G
*************************** 1. row ***************************
              id: 1
     select_type: PRIMARY
           table: <derived2>
      partitions: NULL
            type: system
...
            rows: 1
        filtered: 100.00
           Extra: NULL
*************************** 2. row ***************************
              id: 2
     select_type: DERIVED
           table: NULL
      partitions: NULL
            type: NULL
   possible_keys: NULL
             key: NULL
         key_len: NULL
             ref: NULL
            rows: NULL
        filtered: NULL
           Extra: No tables used
2 rows in set, 1 warning (0.00 sec)
```

15.4 执行计划的 ref 属性

当查询使用索引时,此列表示用于比较的列名或常量(显示 const)。

下面的示例在条件中使用了常量比较。

```
mysql> explain select * from emp where empno = 7369\G
*************************** 1. row ***************************
           id: 1
  select_type: SIMPLE
        table: emp
   partitions: NULL
         type: const
possible_keys: PRIMARY
          key: PRIMARY
      key_len: 2
          ref: const
         rows: 1
     filtered: 100.00
        Extra: NULL
1 row in set, 1 warning (0.00 sec)
```

下面示例的第 2 行表示连接条件使用的列。

```
mysql> explain select * from emp e, dept d where e.deptno = d.deptno \G
*************************** 1. row ***************************
           id: 1
  select_type: SIMPLE
        table: d
   partitions: NULL
         type: ALL
possible_keys: PRIMARY
          key: NULL
      key_len: NULL
          ref: NULL
         rows: 4
     filtered: 100.00
        Extra: NULL
*************************** 2. row ***************************
           id: 1
  select_type: SIMPLE
        table: e
   partitions: NULL
         type: ref
possible_keys: fk_deptno
          key: fk_deptno
      key_len: 2
          ref: db.d.deptno
         rows: 4
     filtered: 100.00
        Extra: NULL
2 rows in set, 1 warning (0.00 sec)
```

15.5　执行计划的 Extra 属性

此列包含了帮助理解执行计划的一些额外信息。下面说明常见取值。

(1) Using index

查询使用了多列覆盖索引。

下面示例的索引 idx_ename_sal 由 ename 和 sal 列创建，执行计划只扫描了索引。

```
mysql> explain select ename, sal from emp where ename = 'SMITH'\G
*************************** 1. row ***************************
           id: 1
  select_type: SIMPLE
        table: emp
   partitions: NULL
         type: ref
possible_keys: idx_ename_sal
          key: idx_ename_sal
      key_len: 43
          ref: const
         rows: 1
     filtered: 100.00
        Extra: Using index
1 row in set, 1 warning (0.00 sec)
```

(2) Using index condition

查询条件使用了普通索引列，而且查询条件会限定多个值。这多个值中，不一定都在普通索引中。若在普通索引中，则要继续访问主键索引，以读取整行记录；若不在普通索引中，则不用再访问主键索引。

下面的查询条件 where deptno in(10, 20)，限定了两个值 10 或 20。

```
mysql> explain select * from emp where deptno in(10,20)\G
*************************** 1. row ***************************
           id: 1
  select_type: SIMPLE
        table: emp
   partitions: NULL
         type: range
possible_keys: fk_deptno
          key: fk_deptno
      key_len: 2
          ref: NULL
         rows: 6
     filtered: 100.00
        Extra: Using index condition
1 row in set, 1 warning (0.00 sec)
```

下面的查询条件 where deptno > 10,以不等式限定了多个值。

```
mysql> explain select * from emp where deptno > 10\G
*************************** 1. row ***************************
           id: 1
  select_type: SIMPLE
        table: emp
   partitions: NULL
         type: range
possible_keys: fk_deptno
          key: fk_deptno
      key_len: 2
          ref: NULL
         rows: 9
     filtered: 100.00
        Extra: Using index condition
1 row in set, 1 warning (0.00 sec)
```

(3) Using where

查询条件中的列未创建索引,使用了全表扫描。

下面示例的 job 列上未创建索引。

```
mysql> explain select * from emp where job = 'CLERK'\G
*************************** 1. row ***************************
           id: 1
  select_type: SIMPLE
        table: emp
   partitions: NULL
         type: ALL
...
         rows: 12
     filtered: 10.00
        Extra: Using where
1 row in set, 1 warning (0.00 sec)
```

(4) Using join buffer (hash join)

当连接查询没有索引可用时,MySQL 在 join buffer 内存缓冲区中执行连接操作。遇到这种情况,应检查是否需要在连接条件列上创建索引。

先删除 emp 表的 deptno 上的外键约束及索引,再删除 dept 表 deptno 列的主键约束,使两个表的 deptno 列都不再存在索引。

```
mysql> alter table emp drop constraint fk_deptno;
mysql> drop index fk_deptno on emp;
mysql> alter table dept drop constraint `PRIMARY`;
```

两个表的连接字段都无索引后,执行下面的命令查看连接查询的执行计划。

```
mysql> explain select e.ename, d.dname from emp e, dept d
    -> where e.deptno = d.deptno \G
*************************** 1. row ***************************
           id: 1
  select_type: SIMPLE
        table: d
   partitions: NULL
         type: ALL
...
         rows: 4
     filtered: 100.00
        Extra: NULL
*************************** 2. row ***************************
           id: 1
  select_type: SIMPLE
        table: e
   partitions: NULL
         type: ALL
...
         rows: 12
     filtered: 10.00
        Extra: Using where; Using join buffer (hash join)
2 rows in set, 1 warning (0.00 sec)
```

(5) Using filesort

出现此值表示排序操作未使用索引,需要使用内存的排序缓冲区执行排序。

```
mysql> explain select ename, sal from emp order by sal\G
*************************** 1. row ***************************
           id: 1
  select_type: SIMPLE
        table: emp
   partitions: NULL
         type: ALL
...
         rows: 12
     filtered: 100.00
        Extra: Using filesort
1 row in set, 1 warning (0.00 sec)
```

(6) Using temporary

在分组查询或排序操作中若出现此值,意味着需要创建临时表保存结果。

```
mysql> explain select job, count(*) from emp
    -> group by job \G
*************************** 1. row ***************************
           id: 1
```

```
    select_type: SIMPLE
         table: emp
    partitions: NULL
          type: ALL
...
          rows: 12
      filtered: 100.00
         Extra: Using temporary
1 row in set, 1 warning (0.00 sec)
```

第16章 统 计 信 息

统计信息是数据的存储和分布状态,准确、及时的统计信息有助于 MySQL 制订有效率的执行计划。统计信息包括两类,一是表或索引的数据分布状态,一般直接称为统计信息(statistics);二是列值的数据分布状态,称为直方图(histogram),直方图是 MySQL 8.0 新加入的功能。

本章主要内容包括:
- 统计信息的内容;
- 统计信息的分类和收集;
- 统计信息的存储和监控;
- 统计信息的更新;
- 直方图的分类和使用。

16.1 统计信息的内容

MySQL 的行都是以主键索引的形式存储,其统计信息即索引的统计信息。索引统计信息主要包括两种,一是索引列值的不重复数,二是索引占用的数据页数。主键索引列值的不重复数即表的总行数。

16.2 统计信息的分类和收集

统计信息分为永久和临时两类,永久统计信息存储于 mysql 系统库的两个数据字典表中,临时统计信息存储于内存结构。永久统计信息由 MySQL 5.6 引入。

MySQL 使用以上哪类统计信息由 innodb_stats_persistent 参数指定,其取值为 0 或 1,默认为 1,即使用永久统计信息,为 0 则使用临时统计信息。

16.2.1 永久统计信息

永久统计信息存储于 mysql 系统库的两组表中。

(1) index_stats 和 table_stats

这两个表是数据字典表,支持多种存储引擎,但用户不能对其执行查询。

(2) innodb_index_stats 和 innodb_table_stats

这两个表是系统表,只支持 InnoDB 存储引擎,用户可以对其执行查询。

为了避免影响效率,收集统计信息时,并不读取整个索引,而是随机读取若干个索引数据页,然后以此为基准,推测整个索引的数据分布情况,因此,统计信息与实际数据相比会有误差。如果索引数据页数尚不足样本页数,则这些数据页都会被扫描,统计信息也是准确的。

与永久统计信息相关的系统参数皆为全局参数,一般都有对应的表级参数,主要包括以下几个。

(1) innodb_stats_persistent

设置是否使用永久统计信息,取值为 0 或 1,默认为 1,即统计信息使用永久方式存储,这也是 MySQL 推荐的方式。下面命令设置 innodb_stats_persistent 为 1。

```
mysql> set persist innodb_stats_persistent = 1;
```

可以在表级指定 stats_persistent 属性为 0 或 1,使其避免使用系统级设置。

下面命令把 emp 表的 stats_persistent 属性设置为 0。

```
mysql> alter table emp stats_persistent = 0;
```

(2) innodb_stats_persistent_sample_pages

设置收集统计信息时,随机读取的数据页页数,默认为 20。

可以根据具体情况,在表级使用 stats_sample_pages 属性设置其随机读取的数据页数。

要注意,上述参数设置的数据页数是针对索引中的每个列的,如果一个索引由 3 个列构成,随机读取数据页页数设置为 20,则需要读取的数据页总数为 $3 \times 20 = 60$。

设置 innodb_stats_persistent_sample_pages 为 30。

```
mysql> set persist innodb_stats_persistent_sample_pages = 30;
```

设置 emp 表的 stats_sample_pages 属性为 10。

```
mysql> alter table emp stats_sample_pages = 10;
```

(3) innodb_stats_include_delete_marked

收集统计信息时,是否包括标记为删除但尚未提交的行,默认为 off。此参数没有表级的对应属性。开启此参数的实质是,收集统计信息时使用 read uncommitted 隔离级别。

(4) innodb_stats_auto_recalc

更新 10% 的行后,是否自动更新统计信息,默认为 on。为了避免小表频繁更新统计信息,此参数开启后,还要满足两次更新间隔不低于 10 s 的要求,这样才会实施统计信息更新。此参数对应的表级属性为 stats_auto_recalc。

开启 emp 表的 stats_auto_recalc 属性。

```
mysql> alter table emp stats_auto_recalc = 1;
```

可以用一个命令同时设置表级的以下几个属性。

```
mysql> alter table emp
    -> stats_persistent = 1,
    -> stats_sample_pages = 10,
    -> stats_auto_recalc = 1;
```

16.2.2 临时统计信息

临时统计信息保存在内存中,其缺点是服务器重启后,统计信息会丢失,需要再次收集。

第一次打开表或 1/16 的行更新时(间隔不少于 16 次更新),会自动收集。收集临时统计信息不能使用后台进程,会影响服务器的运行效率。

除了 16.2.1 节提到的 innodb_stats_persistent 及其对应的表级 stats_persistent 参数外,以下两个系统级参数与临时统计信息相关,这两个参数没有对应的表级参数。

(1) innodb_stats_transient_sample_pages

收集临时统计信息时,随机读取的样本数据页数,默认为 8。

(2) innodb_stats_on_metadata

查询临时统计信息时,是否同时对其更新,默认为 off。查询 information_schema 系统库中的 tables 和 statistics 视图,或执行相关 show 命令,如 show index from、show table status,都属于查询统计信息。

16.3 统计信息的存储和监控

InnoDB 引擎的永久统计信息存储于 mysql 系统库的两个系统表 innodb_index_stats 和 innodb_table_stats 中,具备相关权限的用户(如 root 用户)可以访问这两个系统表得到统计信息。这两个表可以手工修改,以反映数据的最新分布情况,或进行执行计划的测试。

以上统计数据会加载至 mysql 系统库的两个数据字典表 index_stats 和 table_stats,专门用于给用户查询。这两个表不可直接访问,而要通过 show 命令或系统库 information_schema 中的 statistics 和 tables 视图间接访问。系统参数 information_schema_stats_expiry 用于设置 index_stats 和 table_stats 内的统计数据的有效时长,即何时需要重新加载。

16.3.1 innodb_index_stats 系统表

innodb_index_stats 存储所有索引的统计信息。下面通过实例考查其用法。

在 deptno 和 ename 列上创建复合索引 idx_deptno_ename。

```
mysql> create index idx_deptno_ename on emp(deptno, ename);
```

查询 innodb_index_stats,得到 idx_deptno_ename 索引的统计信息。

```
mysql> select index_name, stat_name, stat_value, stat_description
    -> from mysql.innodb_index_stats
    -> where database_name = 'db' and table_name = 'emp'
    -> and index_name = 'idx_deptno_ename';
+------------------+--------------+------------+-----------------------------------+
| index_name       | stat_name    | stat_value | stat_description                  |
+------------------+--------------+------------+-----------------------------------+
| idx_deptno_ename | n_diff_pfx01 |          3 | deptno                            |
| idx_deptno_ename | n_diff_pfx02 |         12 | deptno,ename                      |
| idx_deptno_ename | n_diff_pfx03 |         12 | deptno,ename,empno                |
| idx_deptno_ename | n_leaf_pages |          1 | Number of leaf pages in the index |
| idx_deptno_ename | size         |          1 | Number of pages in the index      |
+------------------+--------------+------------+-----------------------------------+
5 rows in set (0.00 sec)
```

stat_name 和 stat_value 列分别表示统计名称和统计值。

stat_name 的可能取值主要包括以下几种。

(1) n_diff_pfxmm

索引的前 mm 个列的不重复值个数(mm 从 1 开始),在 stat_description 列中包含所描述的前 mm 个列名。

(2) n_leaf_pages

叶节点层的数据页数量。

(3) size

索引的所有数据页页数,包括叶节点和分支节点。

16.3.2 innodb_table_stats 系统表

innodb_table_stats 存储表的统计信息,主要存储内容如下。

- n_rows:表的行数,即主键列值个数。
- clustered_index_size:主键索引的数据页数。
- sum_of_other_index_sizes:所有辅助索引的数据页总数。

下面命令查询 emp 表的统计信息。

```
mysql> select * from mysql.innodb_table_stats
    -> where database_name = 'db' and table_name = 'emp'
    -> \G
*************************** 1. row ***************************
           database_name: db
              table_name: emp
             last_update: 2020-11-14 20:47:11
                  n_rows: 12
    clustered_index_size: 1
sum_of_other_index_sizes: 0
1 row in set (0.00 sec)
```

16.3.3 information_schema 中的统计信息视图

information_schema 系统库中,与统计信息相关的对象主要有 statistics 视图和 tables 视图。

statistics 视图和 tables 视图的查询结果来自 mysql 系统库的数据字典表 index_stats 和 table_stats。index_stats 和 table_stats 中的内容加载自 mysql 系统库的 innodb_index_stats 和 innodb_table_stats,默认超过 24 小时会失效。内容失效,或手工执行 analyze table 命令时,index_stats 和 table_stats 中的统计信息才会重新加载,因此,这些查询结果与实际统计信息相比,一般会有延迟。若要与实际统计信息一致,可把 information_schema_stats_expiry 设置为 0,这种情况下,查询 statistics 视图和 tables 视图,会直接访问 mysql.innodb_index_stats 表和 mysql.innodb_table_stats 表。

下面分别说明以上视图的结构。

statistics 视图中,与统计信息相关的列为 cardinality。除了 cardinality,还包含其他索引相关系统信息,如索引可见性、函数索引的表达式等属性。

查看 statistics 视图的定义可以得知其统计信息来自 mysql.index_stats(略去查询结果)。

```
mysql> show create view information_schema.statistics\G
```

其各列意义如表 16-1 所示。

表 16-1 statistics 视图各列的意义

列名	意义
table_catalog	总为 def
table_schema	表所属的数据库
table_name	表名
non_unique	列值是否重复
index_schema	索引所属的数据库,与 table_schema 相同
index_name	索引名称
seq_in_index	列在索引中的位置,单列索引总为 1
column_name	列名
collation	索引列如何排序,可能取值为 A(升序),D(降序),Null(无序)
cardinality	不重复列值的个数
sub_part	字符串列被索引的字节数,若整列被索引,则为 null
packed	对于 InnoDB 引擎,此列总为 null
nullable	列值可否为 null
index_type	索引类型,如 BTREE
comment	对于 InnoDB 引擎,此列总为 null
index_comment	添加索引时,附加的注释
is_visible	索引是否可见
expression	基于函数索引的函数表达式,普通索引为 null

查询 emp 表上的索引统计信息。

```
mysql> select * from information_schema.statistics
    -> where table_schema = 'db' and table_name = 'emp'
    -> \G
*************************** 1. row ***************************
TABLE_CATALOG: def
 TABLE_SCHEMA: db
   TABLE_NAME: emp
   NON_UNIQUE: 1
 INDEX_SCHEMA: db
   INDEX_NAME: fk_deptno
 SEQ_IN_INDEX: 1
  COLUMN_NAME: deptno
    COLLATION: A
  CARDINALITY: 3
     SUB_PART: NULL
       PACKED: NULL
     NULLABLE: YES
   INDEX_TYPE: BTREE
      COMMENT:
```

```
   INDEX_COMMENT: 
      IS_VISIBLE: YES
     EXPRESSION: NULL
*************************** 2. row ***************************
  TABLE_CATALOG: def
   TABLE_SCHEMA: db
     TABLE_NAME: emp
     NON_UNIQUE: 0
   INDEX_SCHEMA: db
     INDEX_NAME: PRIMARY
   SEQ_IN_INDEX: 1
    COLUMN_NAME: empno
      COLLATION: A
    CARDINALITY: 12
       SUB_PART: NULL
         PACKED: NULL
       NULLABLE: 
     INDEX_TYPE: BTREE
        COMMENT: 
  INDEX_COMMENT: 
      IS_VISIBLE: YES
     EXPRESSION: NULL
2 rows in set (0.00 sec)
```

查看 tables 视图的定义可以得知其统计信息来自 mysql.table_stats（略去查询结果）。

```
mysql> show create view information_schema.tables\G
```

tables 的各列意义如表 16-2 所示。

表 16-2 tables 视图各列的意义

列名	意义
table_catalog	总为 def
table_schema	表所属的数据库
table_name	表名
table_type	表类型，可能取值为 BASE TABLE、VIEW 及 SYSTEM VIEW
engine	表使用的存储引擎名称
version	MySQL 8.0 已不再使用，总为 10
row_format	行的存储格式
table_rows	表的行数，来自主键列的 cardinality
avg_row_length	即 data_length / table_rows
data_length	行数据的大小，即主键索引的页数与页大小的乘积，单位为字节
max_data_length	InnoDB 引擎不适用，总为 null
index_length	普通索引的总大小，即普通索引的数据页总页数与页大小的积
data_free	表所属的表空间中空闲空间的大小

续表

列名	意义
auto_increment	自动增长列的下一个取值
create_time	创建表的时间
update_time	表所在的数据文件最后更新的时间。system 表空间,此值为 null
check_time	表最后执行 check table 的时间
table_collation	字符串排序时,默认使用的排序规则
checksum	InnoDB 不适用,此值总为 null
create_options	建表时附加的选项,如 STATS_AUTO_RECALC
table_comment	建表时附加的注释

查询 emp 表的表数据统计信息。

```
mysql> select * from information_schema.tables
    -> where table_schema = 'db' and table_name = 'emp'
    -> \G
*************************** 1. row ***************************
  TABLE_CATALOG: def
   TABLE_SCHEMA: db
     TABLE_NAME: emp
     TABLE_TYPE: BASE TABLE
         ENGINE: InnoDB
        VERSION: 10
     ROW_FORMAT: Dynamic
     TABLE_ROWS: 12
 AVG_ROW_LENGTH: 1365
    DATA_LENGTH: 16384
MAX_DATA_LENGTH: 0
   INDEX_LENGTH: 16384
      DATA_FREE: 0
 AUTO_INCREMENT: NULL
    CREATE_TIME: 2021-03-02 20:50:45
    UPDATE_TIME: NULL
     CHECK_TIME: NULL
TABLE_COLLATION: utf8mb4_0900_ai_ci
       CHECKSUM: NULL
 CREATE_OPTIONS:
  TABLE_COMMENT:
1 row in set (0.00 sec)
```

16.3.4 show 命令

MySQL 提供了 show index from 和 show table status 命令显示统计信息。show index from 命令的实质是查询视图 information_schema.statistics。show table status 命令的实质是

查询 information_schema.tables 视图。与查询这两个视图相比,因为数据库即当前数据库,所以两个 show 命令不显示表和索引所在的数据库名称。

下面示例使用 show index from 命令查询 emp 表的索引统计信息。

```
mysql> show index from emp\G
*************************** 1. row ***************************
        Table: emp
   Non_unique: 0
     Key_name: PRIMARY
 Seq_in_index: 1
  Column_name: empno
    Collation: A
  Cardinality: 12
     Sub_part: NULL
       Packed: NULL
         Null:
   Index_type: BTREE
      Comment:
Index_comment:
      Visible: YES
   Expression: NULL
*************************** 2. row ***************************
        Table: emp
   Non_unique: 1
     Key_name: fk_deptno
 Seq_in_index: 1
  Column_name: deptno
    Collation: A
  Cardinality: 3
     Sub_part: NULL
       Packed: NULL
         Null: YES
   Index_type: BTREE
      Comment:
Index_comment:
      Visible: YES
   Expression: NULL
2 rows in set (0.01 sec)
```

若 show index from 命令使用 extended 选项,会列出索引包含的所有列。对于普通索引,除了索引列外,也会列出普通索引包含的主键列。对于主键索引,除了主键列外,还会显示 DB_TRX_ID、DB_ROLL_PTR 这两个隐藏列,以及其他所有列,DB_TRX_ID 与 DB_ROLL_PTR 列分别表示正在使用此索引的事务 ID 和指向相应 undo 数据记录的指针。

对 emp 表分别执行以下两个命令(略去显示结果)。

```
mysql> show extended index from emp\G
mysql> show table status like'emp'\G
```

16.3.5 information_schema_stats_expiry 参数

统计信息保存在系统表 mysql.innodb_index_stats 和 mysql.innodb_table_stats 中,为了不影响服务器运行效率,这些数据会再加载至 mysql.index_stats 和 mysql.table_stats 数据字典表,专门用于用户查询。这两个字典表不能直接访问,而要通过 information_schema 系统库中的相关视图或相关 show 命令对其间接访问,如 information_schema 系统库中的 tables、statistics 视图,其定义的基表即这两个数据字典表,而 show index from 及 show table status 命令的查询结果来自以上两个视图。

mysql.innodb_index_stats 和 mysql.innodb_table_stats 中的统计信息自动更新时,数据字典表内的相应数据并不会同步更新,从而会存在一定延迟。

系统参数 information_schema_stats_expiry 用于配置 index_stats 和 table_stats 表内统计信息的有效期,单位为 s,默认为 86 400(即 24 小时),即统计信息加载后,24 小时后会失效。若查询相关统计信息视图或执行相关 show 命令时,字典表的统计数据已失效,则会重新加载。

若 information_schema_stats_expiry 参数设置为 0,则查询统计信息相关视图时,会略过字典表 index_stats 和 table_stats,直接访问系统表 innodb_index_stats 和 innodb_table_stats,从而总是得到最新信息。注意,information_schema_stats_expiry 参数值的改变并不影响系统表内统计信息的更新。

当执行 analyze table 命令,手工更新统计信息时,会把最新的统计信息重新加载至缓冲区,这种情况不受 information_schema_stats_expiry 参数的影响。

16.4 统计信息的更新

统计信息保存在系统表 mysql.innodb_index_stats 和 mysql.innodb_table_stats 中,制订执行计划依据的统计信息即来自这两个表。所谓更新统计信息,是用存储引擎得到的最新统计信息替换这两个表中的旧统计信息。

为了创建正确的执行计划,应该适时更新表的统计信息,使其反映最新的数据分布状态。更新统计信息时,可以配置计划条件,使其自动执行,也可以根据情况,手动执行。

16.4.1 自动更新

若系统或表配置为使用临时统计信息,则会开启自动更新,不需进一步配置。

若系统或表使用永久统计信息,则需要开启系统参数 innodb_stats_auto_recalc,或开启表的 STATS_AUTO_RECALC 选项。表 16-3 所示为两种统计信息在自动更新方面的几个属性的比较。

表 16-3 统计信息的自动更新

属性	永久(persistent)	临时(transient)
触发条件	修改了表中 10% 的行	表的 6.25%

续表

属性	永久（persistent）	临时（transient）
两次更新之间的间隔	至少 10 s	至少 16 次 update 操作
其他因素	无	第 1 次打开表或查询表系统信息
后台进程执行	是	否
系统参数	innodb_stats_auto_recalc	无
表级参数	STATS_AUTO_RECALC	无

16.4.2 手动更新

若统计信息自动更新不满足实际要求，可以根据情况执行手动更新。手动更新使得存储引擎立刻收集指定表的统计信息，更新 innodb_index_stats 和 innodb_table_stats 系统表。MySQL 使用 analyze table 命令和 mysqlcheck 工具手动更新统计信息。

analyze table 命令可以同时更新多个表的统计信息，mysqlcheck 工具可以更新一个或多个数据库内的所有表。

下面的示例为使用 analyze table 命令更新 emp 和 dept 表。

```
mysql> analyze table emp, dept;
```

相比 analyze table 命令，mysqlcheck 工具除了更新统计信息外，还具备其他功能。在更新统计信息方面，mysqlcheck 更加灵活，可以在 shell 程序中执行，也可以利用操作系统工具，如 Linux 的 cron，定时执行。不过，要说明的是，使用 mysqlcheck 更新统计信息时，实质上在背后也是执行 analyze table 命令。

与其他客户端工具相似，使用 mysqlcheck 工具时，需要指定用户名、密码等选项连接至 MySQL 服务器。若更新统计信息，则还需附加--analyze 选项。

使用 mysqlcheck 工具更新统计信息，有以下 3 种方式。

（1）指定一个数据库名称及若干表名

更新统计信息的范围是此数据库内的若干个表，若只指定数据库名称，不指定表名，则更新此数据库内的所有表。

（2）附加--databases 选项

更新统计信息的范围是此选项后指定的所有数据库内的表。

（3）附加--all-databases 选项

更新统计信息的范围包括所有用户数据库及 mysql、sys 系统库内的系统表。

下面用几个实例说明其用法，此处省略了用户名和密码选项。

```
# mysqlcheck --analyze db emp dept
# mysqlcheck --analyze db
# mysqlcheck --analyze --databases db db1
# mysqlcheck --analyze --all-databases
```

16.5 列的直方图

直方图是列值的数据分布状态。设定若干个范围后（每个范围称为一个 bucket，可译为桶),每个范围内的不重复列值个数或列值出现的频率即直方图,列值出现的频率即此范围内的列值个数与表的总行数的百分比。

16.5.1 直方图的作用

直方图可以用来确定某个列值作为条件时,查询结果返回的大致行数,以构造更优的执行计划,如执行表连接查询时,应先访问返回行数较少的表。

在 Oracle 等产品中,直方图也用来在执行计划中决定是否适合使用索引,如果某列值重复率过高(一般超过 20%),则以此值作为查询条件时,可能不适合使用索引,而使用全表扫描。MySQL 的直方图不能用于在执行计划中判断是否使用索引。MySQL 的索引在一定程度上可以代替直方图,对有索引的列,一般不需要再计算直方图。

另外,直方图需要手工执行命令统计,MySQL 不会自动维护。重新计算一个列的直方图,会覆盖之前的直方图数据。

16.5.2 单值直方图和等高直方图

直方图分为单值直方图和等高直方图两种。

单值直方图(singleton):桶个数大于或等于不重复列值的个数。

单值直方图的每个列值对应一个桶。对每个桶,单值直方图用两个值描述,一是桶所对应的列值,二是表示此列值累积频率的浮点数。若 n_1 和 n_2 是排序后的两个相邻列值,n_2 的累积频率减去 n_1 的累积频率即 n_2 占列值总数的百分比。

等高直方图(equi-height):桶个数小于不重复列值的个数。

对每个桶,等高直方图用 4 个值描述:2 个值用来表示桶的上下限,第 3 个值用来表示桶内列值的累积频率,第 4 个值用来表示桶内不重复值的个数。

16.5.3 计算直方图

下面以 emp 表的 mgr 列为例介绍两种直方图。

emp 表的 mgr 列的不重复值有 5 个(不包含 NULL 值),若划分为 5 个桶,则为单值直方图,每个桶对应一个列值。若划分为 3 个桶,则为等高直方图。

计算直方图使用以下命令:

analyze table *table_name* update histogram on *column_name* with *n* buckets;

n 为划分桶的个数,其上限为 1 024。若 *n* 大于不重复列值数量,则桶的实际数量为不重复列值数量。

删除直方图使用以下命令:

analyze table *table_name* drop histogram on *column_name*

查看直方图,可以使用 information_schema 系统库的 column_statistics 视图,其 histogram 列以 JSON 格式显示直方图信息。

下面是 emp 表的 mgr 列上的所有值。

```
mysql> select mgr from emp order by mgr;
+------+
| mgr  |
```

```
+------+
| NULL |
| 7566 |
| 7698 |
| 7698 |
| 7698 |
| 7698 |
| 7698 |
| 7782 |
| 7839 |
| 7839 |
| 7839 |
| 7902 |
+------+
12 rows in set (0.00 sec)
```

手工计算 5 个列值的单值出现频率和累积频率(保留 4 位小数)。

$7566:1/12 = 0.0833$

$7698:5/12 = 0.4167, 0.5$

$7782:1/12 = 0.0833, 0.5833$

$7839:3/12 = 0.2500, 0.8333$

$7902:1/12 = 0.0833, 0.9166$

执行下面命令,计算 mgr 列上的直方图,指定桶个数为 5。

```
mysql> analyze table emp update histogram on mgr with 5 buckets;
```

下面查看直方图的数据情况。

```
mysql> select * from information_schema.column_statistics
    -> where table_name = 'emp'\G
*************************** 1. row ***************************
SCHEMA_NAME: db
 TABLE_NAME: emp
COLUMN_NAME: mgr
  HISTOGRAM: {"buckets": [[7566, 0.08333333333333333], [7698, 0.5], [7782,
              0.5833333333333334], [7839, 0.8333333333333334], [7902, 0.9166666666666667]],
              "data-type": "int", "null-values": 0.08333333333333333, "collation-id": 8,
              "last-updated": "2021-03-03 01:39:44.211347", "sampling-rate": 1.0,
              "histogram-type": "singleton", "number-of-buckets-specified": 5}
1 row in set (0.00 sec)
```

以上结果与手工计算的列值累积频率一致。直方图中包括 null 值的出现频率,但未单独列为一个桶。

执行下面的命令,再次计算 mgr 列上的直方图,指定桶个数为 3。

```
mysql> analyze table emp update histogram on mgr with 3 buckets;
```

下面查看直方图的数据情况。

```
mysql> select * from information_schema.column_statistics
    -> where table_name = 'emp'\G
*************************** 1. row ***************************
SCHEMA_NAME: db
 TABLE_NAME: emp
COLUMN_NAME: mgr
  HISTOGRAM: {"buckets": [[7566, 7698, 0.5, 2], [7782, 7782, 0.5833333333333334, 1], [7839,
7902, 0.9166666666666666, 2]], "data-type": "int", "null-values": 0.08333333333333333, "collation-
id": 8, "last-updated": "2021-03-03 01:44:55.479232", "sampling-rate": 1.0, "histogram-type": "equi-
height", "number-of-buckets-specified": 3}
1 row in set (0.00 sec)
```

每个桶的前两个属性值为桶范围的上下界(桶范围包括这两个值)。

每个桶的第3个属性值为累积出现频率,0.5为满足条件 mgr≤7698 的列值出现的累积频率,0.583 3 为 mgr≤7782 的列值出现的累积频率,0.916 6 为 mgr≤7902 的列值出现的累积频率,计算方法与单值直方图相似,这里不再赘述。

每个桶的第4个属性值为属于相应桶范围的不重复值列值个数。

16.5.4 应用直方图实例

下面使用第 14 章创建的存储过程 big_table 构造两个表,其行数分别为 100 000 和 200 000,并分别命名为 b1 和 b2。

```
mysql> call big_table(100000);
mysql> rename table big_table to b1;
mysql> call big_table(200000);
mysql> rename table big_table to b2;
```

考察下面表连接的执行计划,先访问了 b1 表,再访问 b2 表。其 filtered 列都为 10%,很明显这个数值不准确。满足 id2 = 10 的行,两个表均只有 1 行,准确的 filtered 值应分别为 1/100 000 和 1/200 000,两个值都远远小于执行计划结果中的 10%。

```
mysql> explain select * from b1, b2
    -> where b1.id2 = b2.id2 and b1.id2 = 10
    -> \G
*************************** 1. row ***************************
         id: 1
select_type: SIMPLE
      table: b1
 partitions: NULL
       type: ALL
...
       rows: 99989
   filtered: 10.00
      Extra: Using where
*************************** 2. row ***************************
```

```
         id: 1
  select_type: SIMPLE
        table: b2
   partitions: NULL
         type: ALL
...
         rows: 199672
     filtered: 10.00
        Extra: Using where; Using join buffer (hash join)
2 rows in set, 1 warning (0.01 sec)
```

对 b1 执行 update 操作,将其 50 000 行(即总行数的 50%)的 id2 值改为 10。

```
mysql> update b1 set id2 = 10 where id2 > 50000;
```

重新考查执行计划,可以看到未发生变化。

```
mysql> explain select * from b1, b2 where b1.id2 = b2.id2 and b1.id2 = 10\G
*************************** 1. row ***************************
         id: 1
  select_type: SIMPLE
        table: b1
   partitions: NULL
         type: ALL
...
         rows: 99989
     filtered: 10.00
        Extra: Using where
*************************** 2. row ***************************
         id: 1
  select_type: SIMPLE
        table: b2
   partitions: NULL
         type: ALL
...
         rows: 199672
     filtered: 10.00
        Extra: Using where; Using join buffer (hash join)
2 rows in set, 1 warning (0.00 sec)
```

对 b1 表的 id2 列计算直方图。

```
mysql> analyze table b1 update histogram on id2 with 1024 buckets;
```

再次考查执行计划,可以看到,得知 b1 表的 id2 列值数据分布状态后,访问表的顺序发生了变化,现在先访问了返回行数较少的 b2 表。另外,访问 b1 表的 filtered 列修正为 50,这与满足 id2 = 10 的行占总行数的 50%一致。

```
mysql> explain select * from b1, b2 where b1.id2 = b2.id2 and b1.id2 = 10\G
*************************** 1. row ***************************
         id: 1
  select_type: SIMPLE
        table: b2
   partitions: NULL
         type: ALL
...
```

```
        rows: 199672
    filtered: 10.00
       Extra: Using where
*************************** 2. row ***************************
          id: 1
 select_type: SIMPLE
       table: b1
  partitions: NULL
        type: ALL
...
        rows: 99989
    filtered: 50.00
       Extra: Using where; Using join buffer (hash join)
2 rows in set, 1 warning (0.00 sec)
```

再对 b2 表的 id2 列计算直方图。

```
mysql> analyze table b2 update histogram on id2 with 1024 buckets;
```

考查新的执行计划，访问 b2 表的 filtered 值修正为 0.00。满足 id2 = 10 的记录只有 1 行，准确的 filtered 值为 1 / 2 000 000，若将其商保留两位小数，则为 0.00。现在的执行计划准确地反映了 id2 的分布状态。

```
mysql> explain select * from b1, b2 where b1.id2 = b2.id2 and b1.id2 = 10\G
*************************** 1. row ***************************
          id: 1
 select_type: SIMPLE
       table: b2
  partitions: NULL
        type: ALL
...
        rows: 199672
    filtered: 0.00
       Extra: Using where
*************************** 2. row ***************************
          id: 1
 select_type: SIMPLE
       table: b1
  partitions: NULL
        type: ALL
...
        rows: 99989
    filtered: 50.00
       Extra: Using where; Using join buffer (hash join)
2 rows in set, 1 warning (0.00 sec)
```

第17章 事务处理

事务处理是大型 DBMS 系统软件的重要功能,是数据一致性、并发控制和数据库恢复的基本单位。事务处理是数据库原理的一个重要部分。

本章主要内容包括:
- 事务的概念及应用实例;
- 事务的 ACID 属性;
- 事务隔离级别的概念及设置;
- 事务控制命令;
- 并发控制的 3 个问题。

17.1 事务的概念及应用实例

事务可以看作若干操作的集合,这个集合中的所有操作作为一个整体,要么都完成,要么都取消。数据库应用中涉及事务的实例有很多,下面以银行转账和超市收银为例说明事务的应用场合。

17.1.1 事务应用实例1:银行转账

银行的转账操作是应用事务的一个典型例子。

假定账号 123 要给账号 456 转账 300 元。假定存款余额表为 account,accNo 字段表示账号,balance 字段表示当前余额,这个转账操作可以简化为顺序执行下面两个 update 命令。

```
UPDATE account SET balance = balance - 300 WHERE accNo = 123
UPDATE account SET balance = balance + 300 WHERE accNo = 456
```

若转账操作成功完成,两个账号上的余额之和应该与转账之前相等。如果第 1 个 update 命令执行后,因为计算机故障或停电等原因而导致第 2 个命令不能执行,则两个账号的余额之和会小于转账之前的结果,这在实际应用中是不能接受的。

对于银行转账这类操作,为了避免出现以上问题,我们有必要把以上转账操作的所有命令组成一个事务。如果其中的命令都正常完成,则意味着事务提交成功,即转账操作成功完成;如果在只有一部分完成,而另一部分尚未完成时发生了故障,则应该把完成的部分取消,从而使数据库回到事务开始之前的状态。这样处理,即使出现意外,两个账号的余额之和也会保持不变,不会有人遭受损失。

17.1.2 事务应用实例2:超市收银

再举一个超市收银操作的实例。

假定一顾客买了 2 支钢笔和 10 个笔记本,共 100 元,用银行卡付款。这个付款操作大体包括以下几个步骤：
① 扫描商品条形码，计算总价和商品个数；
② 钢笔库存量－2；
③ 笔记本库存量－10；
④ 顾客银行账号余额－100；
⑤ 超市银行账号余额＋100；
⑥ 给出付款成功提示。

若执行至上面第④步扣款时，银行卡内的余额不够100元，则付款操作会失败，此次购物应取消，扣款操作之前的商品库存量修改操作都应该撤销。很明显，以上这些操作也应该组成一个事务。

17.2 事务的 ACID 属性

计算机应用中的事务应该具备 ACID 属性，ACID 是 4 个英文单词的首字母，这 4 个英文单词是 Atomicity、Consistency、Isolation 和 Durability，分别译为原子性、一致性、隔离性和持久性。下面分别说明其含义。

17.2.1 原子性

原子性指事务中的操作作为一个整体是不可分的，即事务中的所有操作要么都完成，要么都取消。

原子性中的原子是指古希腊哲学家德谟克利特所提出的构成世界万物之最小不可分粒子，而不是现代物理中可以再细分为中子、电子等更小粒子的原子。

一般来说，DBMS 软件本身即可保证事务原子性成立，程序员只需设置事务的边界，即事务的开始和结束。

事务中的操作都完成的标志是事务成功提交，都取消的标志是事务回滚。这两种操作都意味着事务的结束。

17.2.2 一致性

一致性指事务应该把数据库从一个一致状态转到另外一个一致状态。

狭义地说，一致状态是指数据库中的数据都满足表中的约束。

广义地说，一致状态是指数据库中的数据都是正确的，即数据库中的数据应该反映现实情况。因此 C. J. Date 在其《数据库系统引论》一书中，把 ACID 属性中的 C 看作 Correctness。

事务的一致性一般指第 2 种含义，即若事务开始之前，数据库满足正确性要求，则在用户输入数据不存在错误的情况下，经过事务处理后，数据库依然满足正确性要求。要保证事务的正确性往往需要程序员的参与。

如银行转账的实例，作为一个事务，执行至某个中间步骤时，可能数据库中各账户的余额总和与转账前的总额不一致，但整个银行转账事务完成后，数据库中的各账户余额总和应该与转账前相同。

17.2.3 隔离性

隔离性指一个事务对数据修改的结果直到提交后才对其他事务可见。

对于两个不同的事务 A、B，事务 A 可能看到事务 B 的 update 结果（B 提交后），事务 B 也可能看到事务 A 的 update 结果（A 提交后），但两者不可能同时看到对方的 update 结果。

17.2.4 持久性

持久性指事务提交成功后，即使发生了系统故障，其效果在数据库中也是永久的。由于性能的原因，数据并不是在每次事务提交后就写入磁盘。

事务的提交操作会把重做缓冲区的重做数据写入重做文件，若提交后发生了故障，如出现断电或死机等情况，导致修改的数据未写入磁盘，则服务器重启时，会根据重做文件上的重做记录把相关操作再次执行，从而保证数据的持久性。

17.3 事务隔离级别

一个事务与其他事务隔离的程度可以由其隔离级别控制。隔离级别越高，用户越感觉不到其他用户对数据的影响。

SQL 标准指定了以下 4 种隔离级别，MySQL 的默认隔离级别为 repeatable read。

- read uncommitted（未提交读）；
- read committed（提交读）；
- repeatable read（可重复读）；
- serializable（可串行化）。

在 SQL 标准中，使用 set transaction isolation level 命令设置事务的隔离级别，MySQL 一般使用 set transaction_isolation 命令更方便。

17.3.1 read uncommitted

read uncommitted 相当于没有隔离级别。如果一个连接设置为 read uncommitted，则可以读取到其他连接尚未提交的修改结果，即可以查询到事务进行过程中的中间结果。read uncommitted 隔离级别的这种特点破坏了事务的原子性，也破坏了事务的隔离性。

SQL 标准设置此隔离级别的目的是避免读写等待问题。对于某些数据库产品，如 SQL Server，若用户查询的记录正在被另外的事务修改，为了保证事务的原子性，会等待此事务提交后，才读取到结果，这种效果一般通过对读取的数据附加共享锁实现。

MySQL 和 Oracle 等数据库产品使用数据的多版本技术实现读取数据的一致性，如果读取的数据正被其他事务修改，则读取旧版本数据，这样，读取操作一般不需要使用锁，也就不存在读写等待问题，所以 MySQL 和 Oracle 不需要使用 read uncommitted 隔离级别。

17.3.2 read committed

若连接设置为 read committed 隔离级别，则此连接中只能读取到其他事务提交后的结果。

在 MySQL 中，如果一个连接的隔离级别为 read committed，而查询的数据正在被其他事务修改,则此连接不会等待事务提交,而是读取 undo 表空间中的对应旧版本数据,即上一个事务提交后的修改结果。

一份数据存在多个版本的技术称为 MVCC,表示"Multi Version Concurrent Control",当然,在 read committed 隔离级别,一份数据只存在新旧两个版本。

read committed 隔离级别适用于多数应用环境,Oracle 和 SQL Server 默认的隔离级别都为 read committed。

17.3.3 repeatable read

若连接设置为 repeatable read 隔离级别,则在此连接的事务中执行的查询结果,以执行第 1 个查询时的数据状态为基准,即第 1 个查询执行时,其他事务已经提交的修改结果。因而一个事务中的两次相同查询,其结果是相同的,而不管两次查询之间是否有其他事务提交了修改结果。

很明显,在 repeatable read 隔离级别,对于一份数据,不同事务会读取到不同的版本,一份数据的多个旧版本会存储于 undo 表空间。

但是 update 和 delete 操作的搜索范围包括事务开始之后由其他连接提交的修改结果,这种数据查询不到,却可以对其执行 update 或 delete 操作,好像凭空多出来的数据,这种数据一般称为幻象数据(phantom data)。这是 MySQL 数据库可重复读隔离级别的一个特点。MySQL 默认的隔离级别为 repeatable read。

下面说明幻象数据的产生过程。

在连接 1 设置可重复读隔离级别后,查询 emp 表中满足条件 where deptno = 10 的行。

```
conn1> set transaction_isolation = 'repeatable-read';
Query OK, 0 rows affected (0.00 sec)

conn1> start transaction;
Query OK, 0 rows affected (0.00 sec)

conn1> select * from emp where deptno = 10;
+-------+--------+-----------+------+------------+---------+------+--------+
| empno | ename  | job       | mgr  | hiredate   | sal     | comm | deptno |
+-------+--------+-----------+------+------------+---------+------+--------+
|  7782 | CLARK  | MANAGER   | 7839 | 1981-06-09 | 2450.00 | NULL |     10 |
|  7839 | KING   | PRESIDENT | NULL | 1981-11-17 | 5000.00 | NULL |     10 |
|  7934 | MILLER | CLERK     | 7782 | 1982-01-23 | 1300.00 | NULL |     10 |
+-------+--------+-----------+------+------------+---------+------+--------+
3 rows in set (0.00 sec)
```

在连接 2 执行 update 操作,把原属于 20 号部门的 SMITH 的部门改为 10,使其满足连接 1 的查询条件。

```
conn2> update emp set deptno = 10 where ename = 'SMITH';
Query OK, 1 row affected (0.01 sec)
Rows matched: 1  Changed: 1  Warnings: 0
```

回到连接 1,重复之前的查询,可以发现结果依然是之前的 3 行,但以相同的条件执行

update 操作,却发现满足条件的有 4 行,包含了连接 2 修改后的数据。

```
conn1> select * from emp where deptno = 10;
+-------+-------+-----------+------+------------+---------+------+--------+
| empno | ename | job       | mgr  | hiredate   | sal     | comm | deptno |
+-------+-------+-----------+------+------------+---------+------+--------+
|  7782 | CLARK | MANAGER   | 7839 | 1981-06-09 | 2450.00 | NULL |     10 |
|  7839 | KING  | PRESIDENT | NULL | 1981-11-17 | 5000.00 | NULL |     10 |
|  7934 | MILLER| CLERK     | 7782 | 1982-01-23 | 1300.00 | NULL |     10 |
+-------+-------+-----------+------+------------+---------+------+--------+
3 rows in set (0.00 sec)

conn1> update emp set sal = 1000 where deptno = 10;
Query OK, 4 rows affected (0.00 sec)
Rows matched: 4  Changed: 4  Warnings: 0

conn1> select * from emp where deptno = 10;
+-------+-------+-----------+------+------------+---------+------+--------+
| empno | ename | job       | mgr  | hiredate   | sal     | comm | deptno |
+-------+-------+-----------+------+------------+---------+------+--------+
|  7369 | SMITH | CLERK     | 7902 | 1980-12-17 | 1000.00 | NULL |     10 |
|  7782 | CLARK | MANAGER   | 7839 | 1981-06-09 | 1000.00 | NULL |     10 |
|  7839 | KING  | PRESIDENT | NULL | 1981-11-17 | 1000.00 | NULL |     10 |
|  7934 | MILLER| CLERK     | 7782 | 1982-01-23 | 1000.00 | NULL |     10 |
+-------+-------+-----------+------+------------+---------+------+--------+
4 rows in set (0.00 sec)
```

17.3.4 serializable

若连接设置为 serializable 隔离级别,则在此连接中执行的两次相同查询操作得到的结果相同。第 1 次查到的行会被加锁,从而阻挡其他连接改变结果集,因而一个事务内的多次查询结果相同,也不会产生幻象数据。因为读取操作会对记录加锁,所以可能导致读写等待,除非必要,一般不使用 serializable 隔离级别。与 repeatable read 隔离级别相同,serializable 隔离级别也会使用多版本数据。

17.4 事务控制命令

事务控制命令包括设置事务的开始和结束,设置事务的隔离级别以及只读性。

17.4.1 commit 和 rollback 命令

这两个命令的作用是结束事务。

commit 命令以数据的修改永久化结束事务,rollback 命令以撤销数据的全部修改结束事务。commit 一般译为"提交",rollback 一般译为"回滚"。执行 exit 退出 mysql 客户端,MySQL 会自动执行回滚命令。

不同的数据库产品,执行 commit 操作的原理是相似的。

在前面的 ACID 属性中,我们提到,commit 操作并不是把事务中修改的数据写入磁盘永久保存,这是什么原因呢?

假如有 100 个事务对同一个对象进行修改,只需把最后一个事务对数据修改的结果写入磁盘就可以了。如果事务在提交时就把修改结果写入磁盘,显然这 100 个事务会写入磁盘 100 次,效率被无谓地降低了,因此在事务提交时,没必要把修改的数据写入磁盘。另外,为了避免修改大量数据的事务耗费过多内存,在提交之前,有可能已经把修改的数据写入了磁盘,从而可以把部分内存分配给由磁盘新读入的数据。

由以上分析可以看出,事务的提交操作与数据修改结果写入磁盘之间没有关系。既然 commit 操作不会把事务中的修改结果写入磁盘,那么它又做了哪些事情呢?

默认设置下,事务提交后,主要发生了 3 件事:

- 释放事务中产生的锁;
- 把重做缓冲区中的内容写入磁盘的重做文件;
- 把事务提交操作的日志序列号(LSN)写入 InnoDB 引擎的重做文件,以标记数据库处于一致状态的最后时刻。

执行 rollback 操作时,MySQL 使用 undo 表空间的旧版本数据替换被事务修改的数据。rollback 操作完成的时间与其要回滚的数据量相关。

17.4.2 设置事务模式

系统参数 autocommit 用于设置事务模式,默认为 on,即一个命令执行完毕后,自动提交。这种情况下,一个事务只包含一个命令。这种模式称为自动提交。

若把 autocommit 设置为 off 或 0,则用户连接数据库后执行的第 1 个命令即开始了一个事务,直到用户执行 commit 或 rollback 命令结束事务。这种模式称为隐式模式。

在自动提交模式下,即 autocommit 设置为 on 或 1,可以执行 start transaction 命令显式开始一个事务,执行 commit 或 rollback 命令结束事务。这种模式称为显式模式。

在 MySQL 中,begin 或 begin work 是 start transaction 的同义词,start transaction 符合 SQL 标准,推荐使用。

17.4.3 设置事务隔离级别

设置事务隔离级别可以使用专门的命令 set transaction isolation level,也可以使用通用的系统参数设置命令 set。

下面以设置 read committed 隔离级别为例说明各种方法。

使用 global 选项,设置全局范围的隔离级别,下面几个命令等效。

```
mysql> set global transaction isolation level read committed;
mysql> set global transaction_isolation = 'read-committed';
mysql> set @@global.transaction_isolation = 'read-committed';
```

以上设置只对命令执行后的新连接生效,不影响当前连接。

使用 session 选项,设置当前连接的隔离级别。

```
mysql> set session transaction isolation level read committed;
mysql> set session transaction_isolation = 'read-committed';
mysql> set transaction_isolation = 'read-committed';
mysql> set @@session.transaction_isolation = 'read-committed';
```

以上设置对当前连接的所有事务生效,不影响其他连接。若省略生效范围(上面的第 3 个命令),则 set transaction_isolation 的默认选项为 session。

省略 global 和 session 选项,设置当前连接的下一个单个事务的隔离级别。

```
mysql> set transaction isolation level read committed;
mysql> set @@transaction_isolation = 'read-committed';
```

以上设置只影响其后第 1 个事务的隔离级别,从后续第 2 个事务开始,会恢复之前的隔离级别。

对于专门设置隔离级别的 set transaction isolation level 命令,可选的隔离级别名称包括 read uncommitted、read committed、repeatable read、serializable。

对于设置系统参数的 set 命令,隔离级别名称需要使用单引号,且要使用连字符,可选的隔离级别名称包括 read-uncommitted、read-committed、repeatable-read、serializable。

以上诸多命令,一般使用 set transaction_isolation 命令设置当前连接的隔离级别即可满足多数要求,这种方式与 Oracle、SQL Server 等产品使用 set transaction isolation level 命令设置隔离级别的效果相同。如设置 read-committed 隔离级别:

```
mysql> set transaction_isolation = 'read-committed';
```

17.4.4 设置事务只读性

与隔离级别相似的另一个属性是事务的只读性。

事务默认可读写,事务设置为只读后,在其中只能执行查询操作。如果应用场景符合这种要求,设置只读性后,可以提高查询效率。

设置事务只读、可读写可以使用 set transaction 或 set transaction_read_only 命令。推荐使用后者,与 show 命令查询事务只读性时使用的系统参数名称一致。

下面的两个命令分别使用 set transaction 和 set transaction_read_only 命令,设置事务为全局只读,其效果相同。

```
mysql> set global transaction read only;
mysql> set @@global.transaction_read_only = 1;
```

下面的命令重新把事务设置为可读写,两个命令的效果相同。

```
mysql> set global transaction read write;
mysql> set @@global.transaction_read_only = 0;
```

下面的命令把当前连接的事务设置为只读。

```
mysql> set transaction_read_only = 1;
Query OK, 0 rows affected (0.00 sec)
```

设置生效范围与 17.4.3 节设置隔离级别的方法相同,在此不再赘述。

查询事务可读性,执行下面命令。

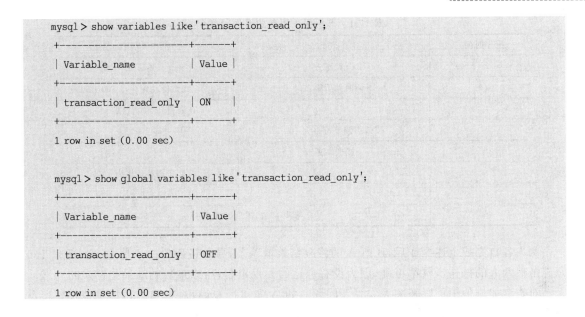

17.5 并发控制要解决的问题

若数据库只由一个用户操作,就不用考虑其他用户的影响,但像 MySQL 这类数据库产品需要对多个用户同时服务。若让多个用户的操作串行化,即一个用户完成其任务后,再开始执行下一个用户的任务,会发生大量的等待情况,这种方式显然不现实。

多个用户同时连接数据库进行操作,如果没有相应的并发控制措施,即使每个用户各自事务中的操作逻辑都是正确的,也会引起很大问题。本节说明由并发操作引起的 3 种问题。

如果没有并发控制,可以导致以下 3 个问题:
- 丢失更新;
- 脏读;
- 不可重复读。

17.5.1 丢失更新

丢失更新是指一个用户对数据的修改结果被另一个用户的修改结果覆盖。

下面以超市修改商品价格为例说明丢失更新是如何发生的。

超市要对某商品降价促销,降价条件有两个:若当前价格高于 100 元时,则把价格调整为原来的 1/2(即 5 折),若库存量高于 1 000 时,则把价格调整为原来的 4/5(即 8 折)。2 个条件是独立的,若 2 个条件都满足,价格 5 折后,再 8 折。如某商品当前价格为 200 元,库存量为 1 500,则降价的 2 个条件均满足,5 折后,其价格降至 100 元,再 8 折后,价格再降至 80 元,最后价格为 80 元。

若工作人员 A 依据第 1 个条件调价,工作人员 B 依据第 2 个条件调价,分别使用 2 个连接,操作顺序如表 17-1 所示。

表 17-1 产生丢失更新的操作顺序

操作时间	工作人员 A	工作人员 B
t_1	开始事务	开始事务
t_2	查询商品价格:200	查询商品库存:1 500
t_3		查询商品当前价格:200
t_4	按 5 折调整价格为 100	
t_5	提交事务	
t_6		按 8 折调整价格为 160
t_7		提交事务
t_8	操作结束,最后价格为 160	

两人各自完成上述操作后,B 把 A 的修改结果覆盖了,导致最后价格不是 80 元,而是 160 元。虽然两人的操作步骤都正确,因为没有合适的并发控制,所以最后的结果是错误的。

对查询加排他锁可以解决丢失更新问题。如 A 先执行查询,对查到的行加排他锁。B 查询价格时,也要加排他锁,与 A 加的锁冲突,需要等待 A 把价格修改完毕提交事务释放锁后,其锁才能加上,这时 B 查到的价格是 A 修改后的结果。B 修改价格时,以 A 修改的结果为基准,从而不会覆盖 A 的结果。

除了 serializable 级别,MySQL 对查询操作默认不使用锁。若有需要,MySQL 支持在 select 命令中使用 for update 子句,对查询到的行加排他锁,效果与执行 update 操作相同。

```
mysql> select * from emp where deptno = 10 for update;
```

17.5.2 脏读

脏读是指一个连接读取了其他尚未提交的事务修改的数据。脏读破坏了事务的原子性,读取到的是事务进行过程中的中间结果,若此结果与事务结束时的结果不同,则此连接读取的是错误数据,如果以此错误数据为依据继续执行另外的任务,则可能造成一连串的错误。

下面以超市收银为例说明脏读的产生过程。

一顾客购买了 2 支钢笔、10 个笔记本,总价 100 元,收银员在连接 A 完成收银操作,超市采购员在连接 B 查询商品库存,确定某种商品是否需要进货。假定收银开始前,钢笔库存量为 100,笔记本库存量为 300,表 17-2 所示是两个连接在不同时间进行的操作。

表 17-2 产生脏读的操作顺序

操作时间	连接 A	连接 B
t_1	开始事务	
t_2	扫描商品条形码,计算总价	
t_3	修改钢笔库存量:100−2 = 98	
t_4	修改笔记本库存量:300−10 = 290	
t_5		查询钢笔库存量:98
t_6		查询笔记本库存量:290
t_7	超市银行账号余额 +100	
t_8	查询顾客银行卡余额小于 100	
t_9	付款失败,回滚事务	

由上述过程可以看到,连接 B 在事务未结束的时候读取了中间结果,导致读取的数据与最终结果不同,若以此数据为依据决定是否进货显然是错误的。

避免把隔离级别设置为 read uncommitted,从而读取未提交的修改结果时,转去读取旧版本数据,即可解决脏读问题。

17.5.3 不可重复读

不可重复读指一个事务中的查询操作因为分为多个步骤,导致其结果包含了另外一个事务开始之前和开始之后的两类数据,从而在最后得到了错误的查询结果。

下面示例中的数据包括 3 个银行账号 acc1、acc2 及 acc3,其余额分别为 100、200、300,如表 17-3 所示。

用户 A 以累加和的方式,计算银行 3 个账号的余额总和。先设置 sum 变量为 0,然后依次查询 3 个账号余额,累加至 sum 变量。

用户 B 执行转账操作,由 acc3 账号转账 100 至 acc1 账号。

两个用户的执行步骤如表 17-4 所示。

表 17-3　各账号初始值

账号	余额
acc1	100
acc2	200
acc3	300
总和	600

表 17-4　产生不可重复读的操作步骤

操作时间	用户 A	用户 B
t_1	开始事务	
t_2	设置 sum=0	
t_3	查询 acc1 账号的余额:100	
t_4	100 累计至 sum;sum=100	
t_5	查询 acc2 账号的余额:200	
t_6	200 累加至 sum;sum=300	
t_7		开始事务
t_8		由 acc3 转账 100 至 acc1
t_9		修改 acc1 账号余额为 200
t_{10}		修改 acc3 账号余额为 200
t_{11}		提交事务
t_{12}	查询 acc3 账号的余额:200	
t_{13}	200 累加至 sum;sum=500	
t_{14}	提交事务	

用户 A 的多次查询操作不存在脏读的情况,但最后得到的总和却是错误的。这里出现错误的原因在于最后的总和包含了用户 B 的事务开始前后的数据。

把事务隔离级别设置为 repeatable read 或 serializable,使用多版本数据技术,即可解决不可重复读问题。

第 18 章 锁

锁是用来控制访问共享资源的一种机制,其目的是把并发操作串行化。数据库系统的一个重要特征是保证多个并发用户一致地读写数据。若并发操作都是写操作,而且是修改同样的数据,就要用到锁的机制了。对于读取操作,MySQL 只在 serializable 隔离级别使用锁。

本章主要内容包括:
- 表锁的各种类型和产生过程;
- 行锁的各种类型和产生过程;
- insert 操作产生的隐式锁;
- 外键对锁的影响;
- 隔离级别及索引对锁的影响;
- 死锁的产生。

18.1 MySQL 的锁类型和锁模式

操作数据库的过程中,用户可以根据需要对记录或表等对象手工加锁,MySQL 也会根据需要对访问资源自动加锁。多数情况下,由 MySQL 自动管理锁就足够了。

下面是 InnoDB 引擎支持的锁类型和锁模式,括号内标注了 performance_schema.data_locks 系统视图的 lock_type 或 lock_mode 列值。

根据锁定类型(lock_type),MySQL 锁可分为
- 表锁(TABLE);
- 行锁(RECORD)。

按照锁定模式(lock_mode),表级锁可分为
- 共享锁(S);
- 排他锁(X);
- 意向共享锁(IS);
- 意向排他锁(IX);
- AUTO_INC 锁。

按照锁定模式(lock_mode),行级锁可分为
- 共享行锁(S, REC_NOT_GAP);
- 排他行锁(X, REC_NOT_GAP;)
- 共享间隙锁(S, GAP);
- 排他间隙锁(X, GAP);
- next-key 共享锁(S);
- next-key 排他锁(X)。

- insert intention 锁(X，GAP，INSERT_INTENTION)。

获得以上所列的锁信息可以使用 performance_schema.data_locks 视图。除了以上所列，还有属于服务器层的元数据锁，要使用 performance_schema.metadata_locks 视图查询元数据锁的信息。下面说明各种锁的含义，并通过实例演示各种锁的产生。

18.2 表 锁

MySQL 的表锁包括共享锁(以 S 表示)和排他锁(以 X 表示)，以及意向共享锁(以 IS 表示)和意向排他锁(以 IX 表示)。表级 S 锁表示要对表执行共享访问，不允许其他事务对表执行任何修改，表级 IS 锁表示需要查询表中的记录。表级 X 锁表示需要对表执行独占访问，表级 IX 锁表示需要修改表中记录。以上 4 种表锁的兼容性如表 18-1 所示。

表 18-1　表锁兼容性

锁模式	X	S	IX	IS
X	×	×	×	×
S	×	√	×	√
IX	×	×	√	√
IS	×	√	√	√

18.2.1　表级 S 锁和表级 X 锁

相对其他锁，表级 S 锁和表级 X 锁比较特殊，实际生产环境较少用到。

这两种锁不属于 InnoDB 存储引擎，而是 server 层的功能，需要执行 lock tables 命令手工对表附加，但 S 和 X 的叫法是 InnoDB 引擎的表示方式。lock tables 命令属于锁管理命令，执行时，会先发出 commit 命令提交之前的事务。

lock tables 命令附加的这两种表级锁需要执行 unlock tables 命令释放，结束事务的 commit 和 rollback 命令不影响这种锁的释放。

使用 lock tables 命令分别对 emp 表和 dept 表附加 S 锁和 X 锁。

```
mysql> lock tables db.emp read, db.dept write;
```

performance_schema.metadata_locks 视图的 lock_type 列中，SHARED_NO_READ_WRITE 表示表级 X 锁，SHARED_READ_ONLY 表示表级 S 锁，这是服务器层的表示方式。

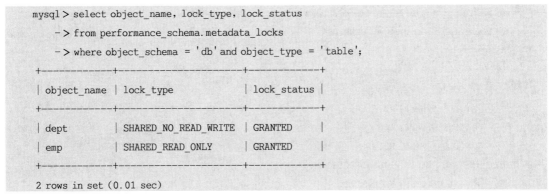

执行 lock tables 命令的前后，会各发出一次 commit 命令，其自身构成的事务就自动提交了。事务和锁的相关视图 information_schema.innodb_trx 及 performance_schema.data_locks 中，不会有 lock tables 命令产生的锁信息。

若要使用系统视图 performance_schema.data_locks 得到 lock tables 命令产生的锁信息，可以把 autocommit 系统变量设置为 0，避免 MySQL 自动提交事务。

```
mysql> set autocommit = 0;
Query OK, 0 rows affected (0.00 sec)

mysql> lock tables emp read, dept write;
Query OK, 0 rows affected (0.00 sec)

mysql> select object_name, lock_type, lock_mode, lock_status
    -> from performance_schema.data_locks;
+-------------+-----------+-----------+-------------+
| object_name | lock_type | lock_mode | lock_status |
+-------------+-----------+-----------+-------------+
| emp         | TABLE     | S         | GRANTED     |
| dept        | TABLE     | X         | GRANTED     |
+-------------+-----------+-----------+-------------+
2 rows in set (0.00 sec)
```

18.2.2 DDL 语句产生的元数据锁

对表执行 DDL 操作时，会产生元数据锁。

在连接 1 对 emp 表执行 update 操作。

```
conn1> start transaction;
Query OK, 0 rows affected (0.00 sec)

conn1> update emp set sal = 880 where ename = 'SMITH';
Query OK, 1 row affected (0.00 sec)
Rows matched: 1  Changed: 1  Warnings: 0
```

在连接 2 创建索引，即执行 DDL 语句，需要获得元数据锁而发生等待。

```
conn2> create index idx_sal on emp(sal);
```

回到连接 1，使用 performance_schema.metadata_locks 查询元数据锁的情况。

```
mysql> select * from performance_schema.metadata_locks
    -> where lock_status = 'PENDING'
    -> \G
*************************** 1. row ***************************
          OBJECT_TYPE: TABLE
        OBJECT_SCHEMA: db
          OBJECT_NAME: emp
          COLUMN_NAME: NULL
OBJECT_INSTANCE_BEGIN: 140555892591632
            LOCK_TYPE: EXCLUSIVE
```

```
        LOCK_DURATION: TRANSACTION
          LOCK_STATUS: PENDING
               SOURCE: mdl.cc:3701
      OWNER_THREAD_ID: 48
       OWNER_EVENT_ID: 10
1 row in set (0.00 sec)
```

也可以使用 information_schema.processlist（show processlist 命令的结果即来自此视图，但不能指定列和过滤条件），以及 sys.schema_table_lock_waits 系统视图。

下面使用 information_schema.processlist 系统视图。

```
conn1> select * from information_schema.processlist
    -> where db = 'db' and state like 'Waiting%'
    -> \G
*************************** 1. row ***************************
     ID: 9
   USER: root
   HOST: localhost
     DB: db
COMMAND: Query
   TIME: 1283
  STATE: Waiting for table metadata lock
   INFO: create index idx_sal on emp(sal)
1 row in set (0.00 sec)
```

18.2.3 表级 IS 锁和表级 IX 锁

意向锁的目的是阻挡表级 S 锁和表级 X 锁。意向排他锁（IX 锁）与 S 锁和 X 锁都不兼容，意向共享锁（IS 锁）只与 X 锁不兼容。两种意向锁相互之间及自身之间都是兼容的。

在 serializable 隔离级别对表执行查询时，会对表附加意向共享锁；在所有隔离级别，对表执行 update、delete 或 insert 操作时，会对表附加意向排他锁。

下面示例验证在 serializable 隔离级别执行查询时，对表附加的意向共享锁。

```
mysql> set transaction_isolation = 'serializable';
Query OK, 0 rows affected (0.01 sec)

mysql> start transaction;
Query OK, 0 rows affected (0.00 sec)

mysql> select ename, sal from emp where deptno = 10;
+--------+---------+
| ename  | sal     |
+--------+---------+
| CLARK  | 2450.00 |
| KING   | 5000.00 |
| MILLER | 1300.00 |
+--------+---------+
```

```
3 rows in set (0.00 sec)

mysql> select object_name, index_name, lock_type, lock_mode, lock_data
    -> from performance_schema.data_locks
    -> where lock_type = 'TABLE';
+-------------+------------+-----------+-----------+-----------+
| object_name | index_name | lock_type | lock_mode | lock_data |
+-------------+------------+-----------+-----------+-----------+
| emp         | NULL       | TABLE     | IS        | NULL      |
+-------------+------------+-----------+-----------+-----------+
1 row in set (0.00 sec)
```

下面示例验证在 repeatable read 隔离级别对表执行 update 操作时,附加的意向排他锁。

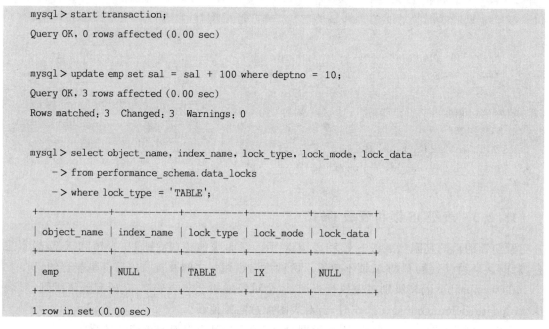

18.3 行 锁

对表执行增删查改时,会对行附加相应行锁。这里的行指主键索引行或普通索引行,主键索引行包括所有列值,即表的行;普通索引行只包括索引列值和主键值。

18.3.1 S,REC_NOT_GAP 锁和 X,REC_NOT_GAP 锁

S,REC_NOT_GAP 锁是对索引行附加的共享锁。X,REC_NOT_GAP 锁是对索引行附加的排他锁,如果索引为主键索引,则锁定的索引行其实就是表的行。

在其他数据库产品中(如 Oracle、SQL Server),排他行锁和共享行锁一般直接用 X 和 S 表示,而 MySQL 的 X 行锁和 S 行锁分别表示排他 next-key 锁和共享 next-key 锁。

在 serializable 隔离级别,使用主键列作为条件执行查询时,会对主键索引行附加共享锁。在所有隔离级别,使用主键列作为条件执行增删改操作时,会对主键索引行附加排他锁。

设置隔离级别为 serializable,以 emp 表的主键列 empno 为条件执行查询,查看此操作产生锁的情况,可以发现对相应记录附加了 S,REC_NOT_GAP 锁。

```
mysql> set transaction_isolation = 'serializable';
Query OK, 0 rows affected (0.00 sec)

mysql> start transaction;
Query OK, 0 rows affected (0.00 sec)

mysql> select ename, sal from emp where empno = 7369;
+-------+--------+
| ename | sal    |
+-------+--------+
| SMITH | 800.00 |
+-------+--------+
1 row in set (0.00 sec)

mysql> select object_name, index_name, lock_type, lock_mode, lock_data
    -> from performance_schema.data_locks;
+-------------+------------+-----------+---------------+-----------+
| object_name | index_name | lock_type | lock_mode     | lock_data |
+-------------+------------+-----------+---------------+-----------+
| emp         | NULL       | TABLE     | IS            | NULL      |
| emp         | PRIMARY    | RECORD    | S,REC_NOT_GAP | 7369      |
+-------------+------------+-----------+---------------+-----------+
2 rows in set (0.00 sec)

mysql> rollback;
Query OK, 0 rows affected (0.01 sec)
```

设置事务隔离级别为 repeatable read,以 emp 表的主键列 empno 为条件执行 update 操作,然后查询锁的情况,可以发现对相应记录附加了 X,REC_NOT_GAP 锁。

```
mysql> set transaction_isolation = 'repeatable-read';
Query OK, 0 rows affected (0.00 sec)

mysql> start transaction;
Query OK, 0 rows affected (0.00 sec)

mysql> update emp set sal = 80 where empno = 7369;
Query OK, 1 row affected (0.00 sec)
Rows matched: 1  Changed: 1  Warnings: 0

mysql> select object_name, index_name, lock_type, lock_mode, lock_status, lock_data
    -> from performance_schema.data_locks;
```

```
+-------------+------------+-----------+---------------+-------------+-----------+
| object_name | index_name | lock_type | lock_mode     | lock_status | lock_data |
+-------------+------------+-----------+---------------+-------------+-----------+
| emp         | NULL       | TABLE     | IX            | GRANTED     | NULL      |
| emp         | PRIMARY    | RECORD    | X,REC_NOT_GAP | GRANTED     | 7369      |
+-------------+------------+-----------+---------------+-------------+-----------+
2 rows in set (0.01 sec)

mysql> rollback;
Query OK, 0 rows affected (0.01 sec)
```

18.3.2 S,GAP 锁和 X,GAP 锁

GAP 锁（一般译为间隙锁），锁定两个值之间的开区间，只用于 repeatable read 和 serializable 两种隔离级别。GAP 锁也分为共享和排他两种，分别表示为 S,GAP 和 X,GAP。使用普通非唯一索引作为条件执行增删查改时，会用到 GAP 锁，目的是禁止其他事务向锁定区间添加记录，这种实现方式比只锁定主键索引行略显笨拙。

对于普通升序索引，GAP 锁的锁定数据（lock_data）在 performance_schema.data_locks 系统视图中会显示为锁定区间的上限值，但实际锁定范围是这个值与比它小的最近相邻值构成的开区间。对于降序索引，锁定数据显示为锁定区间的下限值，但实际锁定范围是这个值与比它大的最近相邻值构成的开区间。

在 serializable 隔离级别执行查询时，若使用非唯一索引搜索行，会使用 S,GAP 锁。

在所有隔离级别执行 update 或 delete 操作时，若使用非唯一索引搜索行，会使用 X,GAP 锁。下面以 serializable 隔离级别为例说明 S,GAP 锁和 X,GAP 锁的产生。

在 ename 列上创建普通索引。

```
mysql> create index idx_ename on emp(ename);
```

将隔离级别设置为 serializable，开始事务后，以 ename 列中不存在的常量值'JOHN'为条件 where ename = 'JOHN'，执行查询。

```
mysql> set transaction_isolation = 'serializable';
Query OK, 0 rows affected (0.00 sec)

mysql> start transaction;
Query OK, 0 rows affected (0.00 sec)

mysql> select * from emp where ename = 'JOHN';
Empty set (0.00 sec)
```

查看锁的情况。

```
mysql> select object_name, index_name, lock_type, lock_mode, lock_data
    -> from performance_schema.data_locks;
+-------------+------------+-----------+-----------+---------------+
| object_name | index_name | lock_type | lock_mode | lock_data     |
+-------------+------------+-----------+-----------+---------------+
| emp         | NULL       | TABLE     | IS        | NULL          |
| emp         | idx_ename  | RECORD    | S,GAP     | 'JONES', 7566 |
+-------------+------------+-----------+-----------+---------------+
```

2 rows in set (0.00 sec)

```
mysql> rollback;
Query OK, 0 rows affected (0.00 sec)
```

在 idx_ename 索引中,与'JOHN'相邻的前后两个值分别为'JAMES'和'JONES',此 GAP 锁的锁定范围是这两个值构成的开区间('JAMES','JONES')。

重新开始事务,再次以条件 where ename = 'JOHN',执行 update,然后查看锁的情况。

```
mysql> start transaction;
Query OK, 0 rows affected (0.00 sec)

mysql> update emp set sal = 800 where ename = 'JOHN';
Query OK, 0 rows affected (0.00 sec)
Rows matched: 0  Changed: 0  Warnings: 0

mysql> select object_name, index_name, lock_type, lock_mode, lock_data
    -> from performance_schema.data_locks;
+-------------+------------+-----------+-----------+---------------+
| object_name | index_name | lock_type | lock_mode | lock_data     |
+-------------+------------+-----------+-----------+---------------+
| emp         | NULL       | TABLE     | IX        | NULL          |
| emp         | idx_ename  | RECORD    | X,GAP     | 'JONES', 7566 |
+-------------+------------+-----------+-----------+---------------+
2 rows in set (0.00 sec)

mysql> rollback;
Query OK, 0 rows affected (0.00 sec)
```

与上面执行查询操作产生锁的情形相似,只是这次的锁模式为 X,GAP。

18.3.3 next-key 锁

next-key 锁也分为共享锁和排他锁,是 GAP 锁与索引行锁的复合,如排他 next-key 锁可看作 X,GAP + X,REC_NOT_GAP,锁定范围是左开右闭区间。

在 performance_schema.data_locks 系统视图的 lock_mode 列中,next-key 共享锁和排他锁分别表示为 S 和 X。

升序索引中,每个数据页中,比此页最大值还大的虚拟记录(可以看作无穷大,$+\infty$),或降序索引中,比最小值还小的虚拟记录(可以看作无穷小,$-\infty$),都称为 supremum pseudo-record;反之,在升序索引中,每个数据页中,比此页最小值还小的虚拟记录,或在降序索引中,比最大值还大的虚拟记录,都称为 infimum pseudo-record。不论升序还是降序,supremum pseudo-record 可以看作每个索引数据页中的最后一个记录,infimum pseudo-record 可以看作每个索引数据页中的第一个记录。

next-key 锁的锁定数据(lock_data)只会用到 supremum pseudo-record。对于升序索引,假设其最大索引列值为 max_value,若锁定数据为 supremum pseudo-record,意味着锁定范围

为(max_value，+∞]；对于降序索引，假设其最小索引列值为 min_value，若锁定数据为 supremum pseudo-record，意味着锁定范围为[-∞，min_value)。-∞和+∞处使用的闭区间符号(方括号)，只是为了看起来更符合 next-key 锁的定义，在数学上，需要表示为开区间(圆括号)。

继续18.3.2节实验过程，以 idx_ename 索引中存在的常量值'SMITH'为条件执行 update 操作，然后查询锁的情况。

```
mysql> set transaction_isolation = 'repeatable-read';
Query OK, 0 rows affected (0.00 sec)

mysql> start transaction;
Query OK, 0 rows affected (0.00 sec)

mysql> update emp set sal = 900 where ename = 'SMITH';
Query OK, 1 row affected (0.01 sec)
Rows matched: 1  Changed: 1  Warnings: 0

mysql> select object_name, index_name, lock_type, lock_mode, lock_data
    -> from performance_schema.data_locks;
+-------------+------------+-----------+---------------+----------------+
| object_name | index_name | lock_type | lock_mode     | lock_data      |
+-------------+------------+-----------+---------------+----------------+
| emp         | NULL       | TABLE     | IX            | NULL           |
| emp         | idx_ename  | RECORD    | X             | 'SMITH', 7369  |
| emp         | PRIMARY    | RECORD    | X,REC_NOT_GAP | 7369           |
| emp         | idx_ename  | RECORD    | X,GAP         | 'TURNER', 7844 |
+-------------+------------+-----------+---------------+----------------+
4 rows in set (0.00 sec)
```

在 idx_ename 索引中，'SMITH'前后的值分别是'MILLER'和'TURNER'。

查询结果的第1行是 update 操作对表附加的意向排他锁。

第2行表示 next-key 排他锁，由索引行排他锁和其前面的 GAP 锁构成，锁定范围为左开右闭区间：('MILLER'，'SMITH']，开区间('MILLER'，'SMITH')为 GAP 锁的锁定区间，lock_data 列值'SMITH'，7369 为锁定的索引行。

第3行表示 update 操作对主键索引记录附加的排他行锁。

第4行表示 update 操作产生的 GAP 锁，其锁定范围是开区间('SMITH'，'TURNER')，不包括上界和下界，lock_data 列值'TURNER'，7844 为索引行。

把 idx_ename 上的两个锁放在一起，锁定范围是'SMITH'所在的索引行及其两侧的开区间。

如果隔离级别设置为 serializable，以相同条件执行查询操作，上面的排他锁会改为使用共享锁，如下面的实验过程所示。

```
mysql> set transaction_isolation = 'serializable';
Query OK, 0 rows affected (0.00 sec)
```

```
mysql> start transaction;
Query OK, 0 rows affected (0.00 sec)

mysql> select * from emp where ename = 'SMITH';
+-------+-------+-------+------+---------------------+--------+------+--------+
| empno | ename | job   | mgr  | hiredate            | sal    | comm | deptno |
+-------+-------+-------+------+---------------------+--------+------+--------+
|  7369 | SMITH | CLERK | 7902 | 1980-12-17 00:00:00 | 800.00 | NULL |     20 |
+-------+-------+-------+------+---------------------+--------+------+--------+
1 row in set (0.00 sec)

mysql> select object_name, index_name, lock_type, lock_mode, lock_data
    -> from performance_schema.data_locks;
+-------------+------------+-----------+---------------+---------------+
| object_name | index_name | lock_type | lock_mode     | lock_data     |
+-------------+------------+-----------+---------------+---------------+
| emp         | NULL       | TABLE     | IS            | NULL          |
| emp         | idx_ename  | RECORD    | S             | 'SMITH', 7369 |
| emp         | PRIMARY    | RECORD    | S,REC_NOT_GAP | 7369          |
| emp         | idx_ename  | RECORD    | S,GAP         | 'TURNER', 7844|
+-------------+------------+-----------+---------------+---------------+
4 rows in set (0.01 sec)
```

下面考查锁定数据为 supremum pseudo-record 的 next-key 锁情形。

在 ename 列上创建升序索引(默认即为升序)。

```
mysql> create index idx_ename on emp(ename asc);
```

设置隔离级别后,执行 update 操作,指定条件为 where ename > 'WARD','WARD'为 ename 列上的最大值。

```
mysql> set transaction_isolation = 'repeatable-read';
Query OK, 0 rows affected (0.00 sec)

mysql> start transaction;
Query OK, 0 rows affected (0.00 sec)

mysql> update emp set sal = 800 where ename > 'WARD';
Query OK, 0 rows affected (0.00 sec)
Rows matched: 0  Changed: 0  Warnings: 0
```

查询锁的情况,可以发现,此时的锁定数据为 supremum pseudo-record,表示锁定范围为 ('WARD', +∞]。

```
mysql> select object_name, index_name, lock_type, lock_mode, lock_data
    -> from performance_schema.data_locks;
+-------------+------------+-----------+-----------+-----------+
| object_name | index_name | lock_type | lock_mode | lock_data |
```

```
+-------------+------------+-----------+-----------+------------------------+
| emp         | NULL       | TABLE     | IX        | NULL                   |
| emp         | idx_ename  | RECORD    | X         | supremum pseudo-record |
+-------------+------------+-----------+-----------+------------------------+
2 rows in set (0.00 sec)

mysql> rollback;
Query OK, 0 rows affected (0.00 sec)
```

回滚以上事务后,在 ename 列上重建降序索引。

```
mysql> drop index idx_ename on emp;
Query OK, 0 rows affected (0.01 sec)
Records: 0  Duplicates: 0  Warnings: 0

mysql> create index idx_ename on emp(ename desc);
Query OK, 0 rows affected (0.01 sec)
Records: 0  Duplicates: 0  Warnings: 0
```

开始事务后,使用条件 where ename < 'ALLEN',执行 update 操作,然后查询锁的情况,可以发现,与上面升序索引的情形相同,锁定数据依然是 supremum pseudo-record。但此时的锁定范围为 $[-\infty, \text{'ALLEN'})$。

```
mysql> start transaction;
Query OK, 0 rows affected (0.00 sec)

mysql> update emp set sal = 800 where ename <'ALLEN';
Query OK, 0 rows affected (0.00 sec)
Rows matched: 0  Changed: 0  Warnings: 0

mysql> select object_name, index_name, lock_type, lock_mode, lock_data
    -> from performance_schema.data_locks;
+-------------+------------+-----------+-----------+------------------------+
| object_name | index_name | lock_type | lock_mode | lock_data              |
+-------------+------------+-----------+-----------+------------------------+
| emp         | NULL       | TABLE     | IX        | NULL                   |
| emp         | idx_ename  | RECORD    | X         | supremum pseudo-record |
+-------------+------------+-----------+-----------+------------------------+
2 rows in set (0.00 sec)
```

18.3.4 insert intention 锁

insert intention 锁是在执行 insert 操作时添加的 X GAP 锁,目的是阻挡向已存在的 GAP 锁锁定区间添加记录。要注意的是,insert intention 锁会等待 X GAP 锁的释放,但 X GAP 锁不必等待 insert intention 锁的释放。

继续 18.3.3 节实验过程,ename 列上已创建索引 idx_ename。

在连接 1,以 where ename = 'SMITH'为条件对 emp 执行 update 操作。

```
conn1> set transaction_isolation = 'repeatable-read';
Query OK, 0 rows affected (0.00 sec)

conn1> start transaction;
Query OK, 0 rows affected (0.01 sec)

conn1> update emp set sal = 880 where ename = 'SMITH';
Query OK, 1 row affected (0.00 sec)
Rows matched: 1  Changed: 1  Warnings: 0
```

查询系统表 performance_schema.data_locks,可以看到以上 update 操作产生的 X GAP 锁,锁定范围为开区间('SMITH','TURNER')。

```
conn1> select thread_id th_id, index_name, lock_type,
    ->        lock_mode, lock_data, lock_status status
    -> from performance_schema.data_locks;
+-------+------------+-----------+-----------------+----------------+---------+
| th_id | index_name | lock_type | lock_mode       | lock_data      | status  |
+-------+------------+-----------+-----------------+----------------+---------+
|    47 | NULL       | TABLE     | IX              | NULL           | GRANTED |
|    47 | idx_ename  | RECORD    | X               | 'SMITH', 7369  | GRANTED |
|    47 | PRIMARY    | RECORD    | X,REC_NOT_GAP   | 7369           | GRANTED |
|    47 | idx_ename  | RECORD    | X,GAP           | 'TURNER', 7844 | GRANTED |
+-------+------------+-----------+-----------------+----------------+---------+
4 rows in set (0.00 sec)
```

在连接 2 对 emp 表添加一行记录,ename 指定为'TRUMP',恰好在('SMITH','TURNER')区间内。

```
conn2> start transaction;
Query OK, 0 rows affected (0.01 sec)

conn2> insert into emp(empno, ename) values(7905,'TRUMP');
```

查看锁的情况,由查询结果的第 2 行可知,以上操作产生了 insert intention 锁,锁定范围为('SMITH','TURNER'),表示要向此范围添加记录,此锁与之前的 X GAP 锁的锁定范围('SMITH','TURNER')不兼容。

```
conn1> select thread_id th_id, index_name, lock_type,
    ->        lock_mode, lock_data, lock_status status
    -> from performance_schema.data_locks;
+-------+------------+-----------+------------------------+----------------+---------+
| th_id | index_name | lock_type | lock_mode              | lock_data      | status  |
+-------+------------+-----------+------------------------+----------------+---------+
|    48 | NULL       | TABLE     | IX                     | NULL           | GRANTED |
|    48 | idx_ename  | RECORD    | X,GAP,INSERT_INTENTION | 'TURNER', 7844 | WAITING |
|    47 | NULL       | TABLE     | IX                     | NULL           | GRANTED |
|    47 | idx_ename  | RECORD    | X                      | 'SMITH', 7369  | GRANTED |
|    47 | PRIMARY    | RECORD    | X,REC_NOT_GAP          | 7369           | GRANTED |
|    47 | idx_ename  | RECORD    | X,GAP                  | 'TURNER', 7844 | GRANTED |
+-------+------------+-----------+------------------------+----------------+---------+
6 rows in set (0.00 sec)
```

若添加记录的 ename 值大于现有最大值(即'WARD'),则其产生的锁模式为 X,INSERT_INTENTION,表示锁定范围为('WARD', supremum pseudo-record]。

为了看到 X,INSERT_INTENTION 锁的产生,回滚连接 1 的操作,对 emp 表执行 update 操作,条件设置为 where ename = 'WARD','WARD'是 ename 列值中最大的一个。

```
conn1> start transaction;
Query OK, 0 rows affected (0.00 sec)

conn1> update emp set sal = 880 where ename = 'WARD';
Query OK, 1 row affected (0.00 sec)
Rows matched: 1  Changed: 1  Warnings: 0

conn1> select thread_id th_id, index_name, lock_type,
    ->        lock_mode, lock_data, lock_status status
    -> from performance_schema.data_locks;
```

th_id	index_name	lock_type	lock_mode	lock_data	status
47	NULL	TABLE	IX	NULL	GRANTED
47	idx_ename	RECORD	X	supremum pseudo-record	GRANTED
47	idx_ename	RECORD	X	'WARD', 7521	GRANTED
47	PRIMARY	RECORD	X,REC_NOT_GAP	7521	GRANTED

4 rows in set (0.00 sec)

以上结果的第 2 行的 X 锁为 next-key 锁,锁定范围为('WARD', supremum pseudo-record]。

在连接 2 添加记录,ename 值指定为'ZENO',使其大于当前最大值'WARD'。

```
conn2> start transaction;
Query OK, 0 rows affected (0.00 sec)

conn2> insert into emp(empno, ename) values(7908,'ZENO');
```

回到连接 1 查看锁的情况。

```
conn1> select thread_id th_id, index_name, lock_type,
    ->        lock_mode, lock_data, lock_status status
    -> from performance_schema.data_locks;
```

th_id	index_name	lock_type	lock_mode	lock_data	status
48	NULL	TABLE	IX	NULL	GRANTED
48	idx_ename	RECORD	X,INSERT_INTENTION	supremum pseudo-record	WAITING
47	NULL	TABLE	IX	NULL	GRANTED
47	idx_ename	RECORD	X	supremum pseudo-record	GRANTED

```
|    47 | idx_ename   | RECORD | X             | 'WARD', 7521 | GRANTED |
|    47 | PRIMARY     | RECORD | X,REC_NOT_GAP | 7521         | GRANTED |
+-------+-------------+--------+---------------+--------------+---------+
6 rows in set (0.00 sec)
```

由以上结果的第 2 行可以看到，本次 insert 操作产生的锁模式为 X, INSERT_INTENTION, 锁定数据为 supremum pseudo-record, 实际的锁定范围是 ('WARD', supremum pseudo-record]。

18.4 insert 操作产生的隐式锁

如果 insert 操作与其他事务的锁不冲突，则 MySQL 不产生相应锁的内存结构，只在其主键索引行和所在数据页写入其事务 ID。如果后续其他事务需要操作此行时，发现此事务尚未结束，则会唤醒相应线程对 insert 操作产生显式锁。如果没有后续事务操作此行，则不需要把锁显式化，从而可以节省内存资源，这种加锁方式称为隐式锁。

在连接 1 执行 insert 操作。此时尚无其他连接，查询锁的情况，可以看到除了表级 IX 锁以外，并没有产生其他锁。

```
conn1> start transaction;
Query OK, 0 rows affected (0.00 sec)

conn1> insert into emp(empno, ename) values(8888,'Mike');
Query OK, 1 row affected (0.00 sec)

conn1> select thread_id th_id, engine_transaction_id tx_id,
    ->        index_name, lock_type, lock_mode, lock_data, lock_status status
    -> from performance_schema.data_locks;
+-------+-------+------------+-----------+-----------+-----------+---------+
| th_id | tx_id | index_name | lock_type | lock_mode | lock_data | status  |
+-------+-------+------------+-----------+-----------+-----------+---------+
|    47 |  8494 | NULL       | TABLE     | IX        | NULL      | GRANTED |
+-------+-------+------------+-----------+-----------+-----------+---------+
1 row in set (0.00 sec)
```

在连接 2 对添加的新记录执行 update 操作，发生等待。

```
conn2> start transaction;
Query OK, 0 rows affected (0.00 sec)

conn2> update emp set sal = 900 where empno = 8888;
```

查询锁的情况，考查下面查询结果的最后一行。

```
conn1> select thread_id th_id, engine_transaction_id tx_id,
    ->        index_name, lock_type, lock_mode, lock_data, lock_status status
    -> from performance_schema.data_locks;
```

```
+-------+-------+------------+-----------+---------------+-----------+---------+
| th_id | tx_id | index_name | lock_type | lock_mode     | lock_data | status  |
+-------+-------+------------+-----------+---------------+-----------+---------+
|    48 |  8495 | NULL       | TABLE     | IX            | NULL      | GRANTED |
|    48 |  8495 | PRIMARY    | RECORD    | X,REC_NOT_GAP | 8888      | WAITING |
|    47 |  8494 | NULL       | TABLE     | IX            | NULL      | GRANTED |
|    48 |  8494 | PRIMARY    | RECORD    | X,REC_NOT_GAP | 8888      | GRANTED |
+-------+-------+------------+-----------+---------------+-----------+---------+
4 rows in set (0.00 sec)
```

其线程 ID 为 48（即连接 2），所在事务 ID 为 8494（连接 1 发起的事务），锁状态为 GRANTED。由以上信息可知，此锁由连接 2 发起，但属于连接 1 的事务，即连接 2 把连接 1 的隐式锁显式化了。然后连接 2 自身再对此索引记录附加排他锁，与以上显式化的排他行锁冲突，从而锁状态为等待（WAITING）。

由于等待超时（默认为 50 s），连接 2 的 update 操作会自动中止。手动回滚事务。

```
conn2> start transaction;
Query OK, 0 rows affected (0.00 sec)


conn2> update emp set sal = 900 where empno = 8888;
ERROR 1205 (HY000): Lock wait timeout exceeded; try restarting transaction
conn2> rollback;
Query OK, 0 rows affected (0.00 sec)
```

再次查看此时锁的情况，被显式化的排他行锁并未释放。

```
conn1> select thread_id th_id, engine_transaction_id tx_id,
    ->        index_name, lock_type, lock_mode, lock_data, lock_status status
    -> from performance_schema.data_locks;
+-------+-------+------------+-----------+---------------+-----------+---------+
| th_id | tx_id | index_name | lock_type | lock_mode     | lock_data | status  |
+-------+-------+------------+-----------+---------------+-----------+---------+
|    47 |  8494 | NULL       | TABLE     | IX            | NULL      | GRANTED |
|    48 |  8494 | PRIMARY    | RECORD    | X,REC_NOT_GAP | 8888      | GRANTED |
+-------+-------+------------+-----------+---------------+-----------+---------+
2 rows in set (0.00 sec)
```

18.5 外键对锁的影响

修改子表的外键值时，除了对子表和子表中影响到的行加锁外，也会对主表被引用的行附加共享锁。

把 emp 的外键 deptno 的值由 10 修改为 20，查询锁的情况，除了对 emp 表附加的锁外，也对 dept 表中的 deptno = 20 的行附加了共享锁，对 dept 表附加了意向共享锁。

```
mysql> start transaction;
Query OK, 0 rows affected (0.00 sec)
```

```
mysql> update emp set deptno = 20 where deptno = 10;
Query OK, 3 rows affected (0.01 sec)

Rows matched: 3  Changed: 3  Warnings: 0
mysql> select object_name, index_name, lock_type, lock_mode, lock_data
    -> from performance_schema.data_locks
    -> order by object_name, index_name;
+-------------+------------+-----------+---------------+-----------+
| object_name | index_name | lock_type | lock_mode     | lock_data |
+-------------+------------+-----------+---------------+-----------+
| dept        | NULL       | TABLE     | IS            | NULL      |
| dept        | PRIMARY    | RECORD    | S,REC_NOT_GAP | 20        |
| emp         | NULL       | TABLE     | IX            | NULL      |
| emp         | fk_deptno  | RECORD    | X             | 10, 7782  |
| emp         | fk_deptno  | RECORD    | X             | 10, 7839  |
| emp         | fk_deptno  | RECORD    | X             | 10, 7934  |
| emp         | fk_deptno  | RECORD    | X,GAP         | 20, 7369  |
| emp         | PRIMARY    | RECORD    | X,REC_NOT_GAP | 7782      |
| emp         | PRIMARY    | RECORD    | X,REC_NOT_GAP | 7839      |
| emp         | PRIMARY    | RECORD    | X,REC_NOT_GAP | 7934      |
+-------------+------------+-----------+---------------+-----------+
10 rows in set (0.00 sec)
```

18.6 不同隔离级别下的加锁方式

事务隔离级别及索引都会影响加锁方式,本节讨论这两种因素对加锁方式的影响。

18.6.1 read uncommitted 与 read committed 隔离级别下的锁

select 操作不使用锁,这与 repeatable read 隔离级别相同。

若条件列未使用索引,或使用了主键索引,则 insert、update、delete 操作对表附加意向排他锁,对行附加排他锁。若无冲突,则 insert 操作对主键行附加隐式排他锁。

若条件列创建了普通非唯一索引,则 insert、update、delete 操作对表附加意向排他锁,对普通索引行和主键行各自附加排他锁。若无冲突,则 insert 操作对主键行附加隐式排他锁。

下面以 read committed 隔离级别为例进行验证。

先验证 update 和 select 操作的加锁特点。

设置连接的隔离级别为 read committed,开始事务后,对 emp 表执行 update 操作,此时条件列 ename 上未创建索引。

```
mysql> set transaction_isolation = 'read-committed';
Query OK, 0 rows affected (0.00 sec)

mysql> start transaction;
Query OK, 0 rows affected (0.00 sec)
```

```
mysql> update emp set sal = 900 where ename = 'SMITH';
Query OK, 1 row affected (0.00 sec)
Rows matched: 1  Changed: 1  Warnings: 0
```

查询此时锁的情况。

```
mysql> select object_name, index_name,
    ->        lock_type, lock_mode, lock_data, lock_status
    -> from performance_schema.data_locks;
+-------------+------------+-----------+-------------+-----------+-------------+
| object_name | index_name | lock_type | lock_mode   | lock_data | lock_status |
+-------------+------------+-----------+-------------+-----------+-------------+
| emp         | NULL       | TABLE     | IX          | NULL      | GRANTED     |
| emp         | PRIMARY    | RECORD    | X,REC_NOT_GAP | 7369    | GRANTED     |
+-------------+------------+-----------+-------------+-----------+-------------+
2 rows in set (0.00 sec)

mysql> rollback;
Query OK, 0 rows affected (0.01 sec)
```

回滚以上操作。

在 ename 列上创建索引,重新执行以上操作,可发现两个索引的相关行附加了排他锁。

```
mysql> create index idx_ename on emp(ename);
Query OK, 0 rows affected (0.03 sec)
Records: 0  Duplicates: 0  Warnings: 0


mysql> start transaction;
Query OK, 0 rows affected (0.00 sec)


mysql> update emp set sal = 900 where ename = 'SMITH';
Query OK, 1 row affected (0.01 sec)
Rows matched: 1  Changed: 1  Warnings: 0


mysql> select object_name, index_name,
    ->        lock_type, lock_mode, lock_data, lock_status
    -> from performance_schema.data_locks;
+-------------+------------+-----------+-------------+----------------+-------------+
| object_name | index_name | lock_type | lock_mode   | lock_data      | lock_status |
+-------------+------------+-----------+-------------+----------------+-------------+
| emp         | NULL       | TABLE     | IX          | NULL           | GRANTED     |
| emp         | idx_ename  | RECORD    | X,REC_NOT_GAP | 'SMITH', 7369 | GRANTED     |
| emp         | PRIMARY    | RECORD    | X,REC_NOT_GAP | 7369          | GRANTED     |
+-------------+------------+-----------+-------------+----------------+-------------+
3 rows in set (0.00 sec)

mysql> rollback;
Query OK, 0 rows affected (0.01 sec)
```

回滚以上操作后,对 emp 表执行 select 操作,条件列依然选择 ename。然后查询锁的情况,可以发现 select 操作未产生锁。

```
mysql> start transaction;
Query OK, 0 rows affected (0.00 sec)

mysql> select * from emp where ename = 'SMITH';
+-------+-------+-------+------+------------+--------+------+--------+
| empno | ename | job   | mgr  | hiredate   | sal    | comm | deptno |
+-------+-------+-------+------+------------+--------+------+--------+
|  7369 | SMITH | CLERK | 7902 | 1980-12-17 | 800.00 | NULL |     20 |
+-------+-------+-------+------+------------+--------+------+--------+
1 row in set (0.00 sec)

mysql> select object_name, index_name,
    ->        lock_type, lock_mode, lock_data, lock_status
    -> from performance_schema.data_locks;
Empty set (0.00 sec)
```

再验证 insert 操作的情形。

保留 ename 上的索引 idx_ename,在连接 1 执行 insert 操作,查询锁的情况,可以发现,此时只对表附加了意向排他锁。

```
conn1> start transaction;
Query OK, 0 rows affected (0.00 sec)


conn1> insert into emp(empno, ename) values(9999,'TRUMP');
Query OK, 1 row affected (0.00 sec)


conn1> select thread_id th_id, engine_transaction_id tx_id,
    ->        index_name, lock_type, lock_mode, lock_data, lock_status status
    -> from performance_schema.data_locks;
+-------+-------+------------+-----------+-----------+-----------+---------+
| th_id | tx_id | index_name | lock_type | lock_mode | lock_data | status  |
+-------+-------+------------+-----------+-----------+-----------+---------+
|    47 |  6201 | NULL       | TABLE     | IX        | NULL      | GRANTED |
+-------+-------+------------+-----------+-----------+-----------+---------+
1 row in set (0.00 sec)
```

在连接 2 对新行执行 update 操作,会发生等待。

```
conn2> start transaction;
Query OK, 0 rows affected (0.00 sec)


conn2> update emp set sal = 800 where ename = 'TRUMP';
```

回到连接 1,再次查看锁的情况。

```
conn1 > select thread_id th_id, engine_transaction_id tx_id,
     ->        index_name, lock_type, lock_mode, lock_data, lock_status status
     -> from performance_schema.data_locks;
+-------+-------+------------+-----------+--------------+--------------+---------+
| th_id | tx_id | index_name | lock_type | lock_mode    | lock_data    | status  |
+-------+-------+------------+-----------+--------------+--------------+---------+
|    48 |  6202 | NULL       | TABLE     | IX           | NULL         | GRANTED |
|    48 |  6202 | idx_ename  | RECORD    | X,REC_NOT_GAP| 'TRUMP', 9999| WAITING |
|    47 |  6201 | NULL       | TABLE     | IX           | NULL         | GRANTED |
|    48 |  6201 | idx_ename  | RECORD    | X,REC_NOT_GAP| 'TRUMP', 9999| GRANTED |
+-------+-------+------------+-----------+--------------+--------------+---------+
4 rows in set (0.00 sec)
```

以上结果的最后一行,表示连接2把属于连接1中的隐式排他锁显式化了。连接2的 thread_id 为 48,连接1的 thread_id 为 47,连接1开启的事务标识为 6201。连接2因为等待超时会自动回滚事务,但连接1中被显式化的锁会一直保持至其事务结束。

若在连接2执行 update 操作时,使用主键列作为条件,则还会把连接1中的主键行排他锁显式化。

在连接2对新行执行 update 操作,发生等待。

```
mysql > start transaction;
Query OK, 0 rows affected (0.00 sec)

mysql > update emp set sal = 880 where empno = 9999;
```

回到连接1,查看此时锁的情况,可以发现,被显式化的锁,除了普通索引行附加的排他锁,还包括主键索引行的排他锁。

```
mysql > select thread_id th_id, engine_transaction_id tx_id,
     ->        index_name, lock_type, lock_mode, lock_data, lock_status status
     -> from performance_schema.data_locks;
+-------+-------+------------+-----------+--------------+--------------+---------+
| th_id | tx_id | index_name | lock_type | lock_mode    | lock_data    | status  |
+-------+-------+------------+-----------+--------------+--------------+---------+
|    48 |  6205 | NULL       | TABLE     | IX           | NULL         | GRANTED |
|    48 |  6205 | PRIMARY    | RECORD    | X,REC_NOT_GAP| 9999         | WAITING |
|    47 |  6201 | NULL       | TABLE     | IX           | NULL         | GRANTED |
|    48 |  6201 | idx_ename  | RECORD    | X,REC_NOT_GAP| 'TRUMP', 9999| GRANTED |
|    48 |  6201 | PRIMARY    | RECORD    | X,REC_NOT_GAP| 9999         | GRANTED |
+-------+-------+------------+-----------+--------------+--------------+---------+
5 rows in set (0.00 sec)
```

再考查 ename 列上不存在索引的情况。

删除 ename 列上的索引 idx_ename 后,对 emp 表执行 insert 操作,此时只能查看到表上的意向排他锁。

```
conn1> drop index idx_ename on emp;
Query OK, 0 rows affected (0.01 sec)
Records: 0  Duplicates: 0  Warnings: 0

conn1> start transaction;
Query OK, 0 rows affected (0.00 sec)

conn1> insert into emp(empno, ename) values(9999,'TRUMP');
Query OK, 1 row affected (0.00 sec)

conn1> select thread_id th_id, engine_transaction_id tx_id,
    ->        index_name, lock_type, lock_mode, lock_data, lock_status status
    -> from performance_schema.data_locks;
+-------+-------+------------+-----------+-----------+-----------+---------+
| th_id | tx_id | index_name | lock_type | lock_mode | lock_data | status  |
+-------+-------+------------+-----------+-----------+-----------+---------+
|    47 |  6213 | NULL       | TABLE     | IX        | NULL      | GRANTED |
+-------+-------+------------+-----------+-----------+-----------+---------+
1 row in set (0.00 sec)
```

在连接 2 执行 update 操作,使用未创建索引的 ename 列作为条件。此操作不会产生等待。

```
conn2> start transaction;
Query OK, 0 rows affected (0.00 sec)

conn2> update emp set sal = 800 where ename = 'TRUMP';
Query OK, 0 rows affected (0.00 sec)
Rows matched: 0  Changed: 0  Warnings: 0
```

回到连接 1 查看此时锁的情况,可以发现,连接 2 的 update 操作只是把连接 1 的隐式锁显式化了。

```
conn1> select thread_id th_id, engine_transaction_id tx_id,
    ->        index_name, lock_type, lock_mode, lock_data, lock_status status
    -> from performance_schema.data_locks;
+-------+-------+------------+-----------+---------------+-----------+---------+
| th_id | tx_id | index_name | lock_type | lock_mode     | lock_data | status  |
+-------+-------+------------+-----------+---------------+-----------+---------+
|    48 |  6214 | NULL       | TABLE     | IX            | NULL      | GRANTED |
|    47 |  6213 | NULL       | TABLE     | IX            | NULL      | GRANTED |
|    48 |  6213 | PRIMARY    | RECORD    | X,REC_NOT_GAP | 9999      | GRANTED |
+-------+-------+------------+-----------+---------------+-----------+---------+
3 rows in set (0.00 sec)
```

若在连接 2 继续执行 update 操作,但本次使用 empno 作为条件,此时会发生等待。

```
conn2> update emp set sal = 880 where empno = 9999;
```

回到连接 1,查询此时锁的情况,会发现,连接 2 会对主键行加排他锁,导致与连接 1 中被显式化的排他锁冲突,从而发生等待。

```
conn1 > select thread_id th_id, engine_transaction_id tx_id,
    ->        index_name, lock_type, lock_mode, lock_data, lock_status status
    -> from performance_schema.data_locks;
+-------+-------+------------+-----------+---------------+----------+---------+
| th_id | tx_id | index_name | lock_type | lock_mode     | lock_data | status  |
+-------+-------+------------+-----------+---------------+----------+---------+
|    48 |  6214 | NULL       | TABLE     | IX            | NULL     | GRANTED |
|    48 |  6214 | PRIMARY    | RECORD    | X,REC_NOT_GAP | 9999     | WAITING |
|    47 |  6213 | NULL       | TABLE     | IX            | NULL     | GRANTED |
|    48 |  6213 | PRIMARY    | RECORD    | X,REC_NOT_GAP | 9999     | GRANTED |
+-------+-------+------------+-----------+---------------+----------+---------+
4 rows in set (0.00 sec)
```

18.6.2 repeatable read 隔离级别下的锁

select 操作不使用锁。

insert、update 及 delete 操作对表附加 IX 锁。

对于 update 及 delete 操作,若主键列(或唯一索引列)作为条件列,则附加行级排他锁,若使用非唯一普通索引列作为条件列,则使用 next-key 或 GAP 排他锁。

在不同场景,insert 操作会产生 insert 意向锁、隐式锁等不同种类的锁。以上相关结论,请参考 18.6.1 节内容。

若条件列无索引,update 及 delete 操作会对所有主键记录附加 next-key 锁,其他连接不能对表执行任何 insert、update、delete 操作。

下面以 update 操作为例进行验证。

执行 update 操作,条件列为无索引的 ename。

```
mysql > set transaction_isolation = 'repeatable-read';
Query OK, 0 rows affected (0.00 sec)

mysql > start transaction;
Query OK, 0 rows affected (0.00 sec)

mysql > update emp set sal = 80 where ename = 'SMITH';
Query OK, 1 row affected (0.00 sec)
Rows matched: 1  Changed: 1  Warnings: 0
```

查询锁的情况,可以发现,在所有 12 行主键索引记录及 supremum 虚拟记录上都附加了 next-key 排他锁(即 X 锁)。

```
mysql > select index_name, lock_type, lock_mode, lock_data, lock_status
    -> from performance_schema.data_locks;
+------------+-----------+-----------+-----------+-------------+
| index_name | lock_type | lock_mode | lock_data | lock_status |
```

```
+---------+--------+----+------------------------+---------+
| NULL    | TABLE  | IX | NULL                   | GRANTED |
| PRIMARY | RECORD | X  | supremum pseudo-record | GRANTED |
| PRIMARY | RECORD | X  | 7369                   | GRANTED |
| PRIMARY | RECORD | X  | 7499                   | GRANTED |
| PRIMARY | RECORD | X  | 7521                   | GRANTED |
| PRIMARY | RECORD | X  | 7566                   | GRANTED |
| PRIMARY | RECORD | X  | 7654                   | GRANTED |
| PRIMARY | RECORD | X  | 7698                   | GRANTED |
| PRIMARY | RECORD | X  | 7782                   | GRANTED |
| PRIMARY | RECORD | X  | 7839                   | GRANTED |
| PRIMARY | RECORD | X  | 7844                   | GRANTED |
| PRIMARY | RECORD | X  | 7900                   | GRANTED |
| PRIMARY | RECORD | X  | 7902                   | GRANTED |
| PRIMARY | RECORD | X  | 7934                   | GRANTED |
+---------+--------+----+------------------------+---------+
14 rows in set (0.00 sec)
```

18.6.3　serializable 隔离级别下的锁

select 操作会使用锁。

insert、update 及 delete 操作的加锁方式与 repeatable-read 隔离级别下的相同,这里不再赘述。

下面考查 select 操作的加锁情况。

对表附加 IS 锁。

对于行锁,则有如下规律:

- 若使用无索引列作为条件列,则对所有主键记录附加 next-key 共享锁。
- 若使用唯一索引列作为条件列,则对符合条件的索引记录附加共享锁。
- 若使用普通非唯一索引列作为条件列,则对符合条件的索引记录附加 next-key 共享锁。

先考查查询操作在条件列无索引的情况下产生锁的情况。

```
mysql> set transaction_isolation = 'serializable';
Query OK, 0 rows affected (0.00 sec)

mysql> start transaction;
Query OK, 0 rows affected (0.01 sec)

mysql> select * from emp where ename = 'SMITH';
+-------+-------+-------+------+------------+--------+------+--------+
| empno | ename | job   | mgr  | hiredate   | sal    | comm | deptno |
+-------+-------+-------+------+------------+--------+------+--------+
|  7369 | SMITH | CLERK | 7902 | 1980-12-17 | 800.00 | NULL |     20 |
+-------+-------+-------+------+------------+--------+------+--------+
1 row in set (0.00 sec)
```

查询锁的情况，可以看到，上面查询操作对所有行附加了 next-key 共享锁。

```
mysql> select index_name, lock_type, lock_mode, lock_data
    -> from performance_schema.data_locks;
+------------+-----------+-----------+------------------------+
| index_name | lock_type | lock_mode | lock_data              |
+------------+-----------+-----------+------------------------+
| NULL       | TABLE     | IS        | NULL                   |
| PRIMARY    | RECORD    | S         | supremum pseudo-record |
| PRIMARY    | RECORD    | S         | 7369                   |
| PRIMARY    | RECORD    | S         | 7499                   |
| PRIMARY    | RECORD    | S         | 7521                   |
| PRIMARY    | RECORD    | S         | 7566                   |
| PRIMARY    | RECORD    | S         | 7654                   |
| PRIMARY    | RECORD    | S         | 7698                   |
| PRIMARY    | RECORD    | S         | 7782                   |
| PRIMARY    | RECORD    | S         | 7839                   |
| PRIMARY    | RECORD    | S         | 7844                   |
| PRIMARY    | RECORD    | S         | 7900                   |
| PRIMARY    | RECORD    | S         | 7902                   |
| PRIMARY    | RECORD    | S         | 7934                   |
+------------+-----------+-----------+------------------------+
14 rows in set (0.00 sec)

mysql> rollback;
Query OK, 0 rows affected (0.00 sec)
```

以主键索引列 empno 为条件列执行查询，则对符合条件的主键记录附加共享锁。

```
mysql> start transaction;
Query OK, 0 rows affected (0.00 sec)

mysql> select * from emp where empno = 7369;
+-------+-------+-------+------+------------+--------+------+--------+
| empno | ename | job   | mgr  | hiredate   | sal    | comm | deptno |
+-------+-------+-------+------+------------+--------+------+--------+
|  7369 | SMITH | CLERK | 7902 | 1980-12-17 | 800.00 | NULL |     20 |
+-------+-------+-------+------+------------+--------+------+--------+
1 row in set (0.00 sec)

mysql> select index_name, lock_type, lock_mode, lock_data
    -> from performance_schema.data_locks;
+------------+-----------+-----------+-----------+
| index_name | lock_type | lock_mode | lock_data |
+------------+-----------+-----------+-----------+
| NULL       | TABLE     | IS        | NULL      |
```

```
| PRIMARY      | RECORD     | S,REC_NOT_GAP  | 7369         |
+--------------+------------+----------------+--------------+
2 rows in set (0.00 sec)

mysql> rollback;
Query OK, 0 rows affected (0.00 sec)
```

在 ename 列创建普通索引后，以 where ename = 'SMITH'为条件执行查询，可以发现使用普通索引列作为查询条件，select 操作与 update 操作产生的锁相似，只是类别由共享锁变为排他锁。

```
mysql> create index idx_ename on emp(ename);
Query OK, 0 rows affected (0.02 sec)
Records: 0  Duplicates: 0  Warnings: 0

mysql> start transaction;
Query OK, 0 rows affected (0.00 sec)

mysql> select * from emp where ename = 'SMITH';
+-------+-------+-------+------+------------+--------+------+--------+
| empno | ename | job   | mgr  | hiredate   | sal    | comm | deptno |
+-------+-------+-------+------+------------+--------+------+--------+
|  7369 | SMITH | CLERK | 7902 | 1980-12-17 | 800.00 | NULL |     20 |
+-------+-------+-------+------+------------+--------+------+--------+
1 row in set (0.00 sec)

mysql> select index_name, lock_type, lock_mode, lock_data
    -> from performance_schema.data_locks;
+------------+-----------+---------------+----------------+
| index_name | lock_type | lock_mode     | lock_data      |
+------------+-----------+---------------+----------------+
| NULL       | TABLE     | IS            | NULL           |
| idx_ename  | RECORD    | S             | 'SMITH', 7369  |
| PRIMARY    | RECORD    | S,REC_NOT_GAP | 7369           |
| idx_ename  | RECORD    | S,GAP         | 'TURNER', 7844 |
+------------+-----------+---------------+----------------+
4 rows in set (0.00 sec)

mysql> rollback;
Query OK, 0 rows affected (0.00 sec)
```

18.7 死　　锁

锁可以把并发修改操作串行化，也可以解决并发操作中的丢失更新问题，但如果锁不能被合理使用，也会引起死锁问题。

死锁是两个或多个事务同时处于等待状态，每个事务都在等待另一个事务释放对某个资源的锁后，才能继续自己的操作的过程。

表 18-2 可以说明死锁的产生过程，表中的 R1 和 R2 分别表示两种需要锁定的资源，加锁时采用排他方式。

表 18-2　产生死锁的操作步骤

时间	事务 A	事务 B
t_1	锁定 R1	
t_2		锁定 R2
t_3	锁定 R2	
t_4	等待	锁定 R1
t_5	等待	等待

下面用实际操作在 MySQL 中模拟死锁的产生过程，假定两个事务需要修改 dept 表的 10 号部门和 20 号部门的地址，很明显，这两行记录相当于表 18-2 中需要锁定的资源 R1 和 R2，锁定方式为排他。

启动两个连接，在连接 1 中修改 dept 表中 10 号部门的地址：

```
conn1 > start transaction;
Query OK, 0 rows affected (0.00 sec)

conn1 > update dept set loc = 'BOSTON' where deptno = 10;
Query OK, 1 row affected (0.01 sec)
Rows matched: 1　Changed: 1　Warnings: 0
```

以上操作对 10 号记录附加了排他锁。

在连接 2 中修改 dept 表中 20 号部门的地址：

```
conn2 > start transaction;
Query OK, 0 rows affected (0.00 sec)

conn2 > update dept set loc = 'CHICAGO' where deptno = 20;
Query OK, 1 row affected (0.00 sec)
Rows matched: 1　Changed: 1　Warnings: 0
```

以上操作对 20 号记录附加了排他锁。

回到连接 1，修改 20 号记录，因为 20 号记录已被连接 2 锁定，连接 1 的本次修改操作发生等待。

```
conn1 > update dept set loc = 'CHICAGO' where deptno = 20;
```

回到连接 2，修改 10 号记录，因为 10 号记录已被连接 1 锁定，连接 2 的本次修改操作发生等待。

```
conn2> update dept set loc = 'BOSTON' where deptno = 10;
ERROR 1213 (40001): Deadlock found when trying to get lock; try restarting transaction
```

两个连接都在等待,这时就发生了死锁。

发生死锁后,对资源锁定的等待形成了一个闭环结构,要解除死锁,只要破坏环形结构中的任意一个环节即可。MySQL 会自动探测到死锁的发生,并选择一个事务作为牺牲品对其执行回滚,从而解除死锁。在以上示例中,MySQL 回滚了连接 2 的事务,并给出了相关错误信息。

第 19 章 备 份 恢 复

随着 MySQL 数据库的功能不断增强,很多涉及金融的关键应用使用了 MySQL 作为数据库服务器。完备的备份恢复能力是企业级数据库服务器必须具备的。

本章主要内容包括:
- 备份种类和工具介绍;
- 导出导入操作;
- mysqldump 和 mysqlpump 执行逻辑备份;
- mysqlbinlog 转换日志文件;
- 基于时点的数据库恢复实例。

19.1 备 份 种 类

企业级数据库产品都会提供备份恢复功能,根据不同需要,执行不同种类的备份恢复操作。本节对备份种类做简单介绍,这些分类对于不同厂商的数据库产品都是相似的。

19.1.1 热备份与冷备份

热备份是指在服务器正常运行时,执行备份。执行热备份时,其他应用可以继续对数据库进行读写操作,必须不间断运行的服务器需要采取这种方式。使用这种方式时,备份的数据存在事务不一致的问题,即可能会备份未结束的事务修改的数据,在恢复时,一般需要执行事务一致性恢复,才能打开数据库。

冷备份是指在服务器正常关闭后,复制数据文件。这种情况下,事务都正常结束了,可以用冷备份直接打开数据库,不需要执行后续的事务一致性恢复。

19.1.2 逻辑备份与物理备份

逻辑备份的内容为表的行数据或恢复数据的 SQL 语句,如 create table 及 insert 语句。
物理备份的内容为数据文件或非空数据页。

19.1.3 全备份与增量备份

全备份是备份整个数据库的数据,对于 MySQL 来说,全备份的内容一般为整个实例中的所有数据库。增量备份(incremental backup)的内容为上次备份操作以来改变后的数据页。

19.2 恢 复 种 类

数据丢失或数据库出现故障时,需要对其恢复。根据存储介质是否损坏可以分为介质恢复和实例恢复。

若磁盘等存储介质损坏,则需要使用之前的数据库备份恢复数据库,整个操作过程需要数据库管理员的参与。

若数据库正在运行时,服务器发生了故障(死机、断电等问题),导致数据库意外关闭,则在问题解决、数据库重启时,需要进行实例恢复。实例恢复包括前滚和回滚两个过程,整个过程由 MySQL 自动完成,不需要用户参与。

由事务相关内容我们看到,commit 操作与事务中被修改的数据写入磁盘没有关系,也就是说,在事务提交之前,其修改的部分数据可能已经写入磁盘,而提交之后,可能会有部分被修改的数据尚未写入磁盘,这样的数据库显然处于不一致的状态。

checkpoint 线程启动时,会把数据缓冲区中的脏块(即修改过的数据块)写入磁盘的数据文件,并在 redo log 文件中记录下其 LSN(Log Sequence Number,日志序列号)。最后一次 checkpoint 写入的 LSN 是实例恢复的起点,因为之前的脏块已经被写入磁盘了。

实例恢复开始时,MySQL 读取重做日志文件中最后一次 checkpoint 之后的所有重做数据,然后应用到数据文件,这个步骤称为前滚。

前滚完成后,对于最后一次 checkpoint 之后提交的事务来说,其修改的数据全部写入了磁盘,保证了这些数据的一致性。除了这些提交的事务,还存在着 checkpoint 之后未结束的事务,这些事务中修改的数据及其对应的旧版本数据由于前滚操作,也被分别写入了磁盘中的数据文件及 undo 数据文件,这些数据显然不符合一致性要求。

回滚操作是把最后一次 checkpoint 之后未结束事务的修改效果撤销,即用 undo 表空间中的旧版本数据替换数据文件中被修改的数据。

前滚和回滚完成后,实例恢复即完成,数据库重新处于一致状态。

19.3 备份恢复工具

社区版的 MySQL 只能执行逻辑备份,下面介绍几个常用的工具及 SQL 命令。

mysqldump 及 mysqlpump 工具产生重建数据库及表的 SQL 命令。恢复数据库时,在 mysql 工具中执行这些 SQL 命令即可。

mysqlbinlog 工具把 binary log 文件转换为 SQL 命令。执行全库恢复后,选择合适的起点,继续执行这些命令,可以把数据库恢复至出现故障的时刻。

使用 mysqldump、mysqlpump 及 mysqlbinlog 这几个工具时,均需连接至 MySQL 服务器,本章以用户配置文件存储用户名及密码(/root/.my.cnf),示例中不需使用-u 和-p 选项指定用户名和密码。

select into outfile 命令及 mysqldump(使用--tab 选项)工具可以把表的行导出为指定格式的文本文件。导入这类文本文件,可以使用 mysqlimport 工具或 load data infile 命令。

MySQL Enterprise Backup(mysqlbackup)是 MySQL 官方提供的备份恢复工具,可以执行物理备份,也可以执行增量备份,但此工具非开源,使用时,需付费得到授权。物理备份的内容为数据文件的复制,增量备份的内容为上次备份以来改变过的数据页。第三方的开源免费备份恢复工具 xtrabackup 的功能与 MySQL Enterprise Backup 相似。

19.4 导出导入数据行

导出操作把指定表的行或表的查询结果转换为文本文件。多数数据库产品都会提供文本形式的导入导出工具，可以作为表数据的备份，或在自身产品之间或不同数据库产品之间传递数据。select into outfile 命令与 mysqldump 工具执行导出操作，load data infile 命令和 mysqlimport 工具执行导入操作。

select into outfile 属于 SQL 命令，用于导出查询结果的行，在 mysql 客户端中执行。mysqldump 是独立的客户端工具，除了导出表的行外，还同时导出建表语句。两个导出方式只是使用方法不同，产生的数据行导出文件没有区别。

load data infile 属于 SQL 命令，在 mysql 客户端中执行，导入数据时，需要指定表名。mysqlimport 是独立的客户端工具，导入数据时，要指定数据库名称，不能指定表名，其导入的目标表名称由导出文件的文件名指定。除了以上所列，两个导入方法没有其他区别。

19.4.1 一个简单示例

如果导出文件不使用分隔符，导出、导入操作很简单，下面以 select into outfile 及 load data 为例说明其使用步骤。

导出 db 数据库的 emp 表。

```
mysql> select * from db.emp
    -> into outfile '/var/lib/mysql-files/emp.dmp';
Query OK, 12 rows affected (0.01 sec)
```

清空 emp 表。

```
mysql> delete from db.emp;
Query OK, 12 rows affected (0.00 sec)
```

把导出数据导入 emp 表。

```
mysql> load data infile '/var/lib/mysql-files/emp.dmp'
    -> into table db.emp;
Query OK, 12 rows affected (0.02 sec)
Records: 12  Deleted: 0  Skipped: 0  Warnings: 0
```

默认情况下，导出文件的列值不使用特殊字符括住，列值之间使用 tab 字符分隔，null 值使用"\N"标识，行使用"\n"作为结束标记。

下面是导出文件的内容。

```
[root /var/lib/mysql-files 2021-03-03 17:03:58]
# cat emp.dmp
7369    SMITH   CLERK       7902    1980-12-17   800.00   \N       20
7499    ALLEN   SALESMAN    7698    1981-02-02  1600.00  300.00    30
7521    WARD    SALESMAN    7698    1981-02-22  1250.00  500.00    30
7566    JONES   MANAGER     7839    1981-04-02  2975.00  \N        20
...
7934    MILLER  CLERK       7782    1982-01-23  1300.00  \N        10
```

19.4.2 使用 select into outfile 及 load data 执行导出导入

下面示例使用 select into outfile 导出 emp 表的数据，并指定导出数据的各种分隔符。

```
mysql> select ename, hiredate, sal, comm from emp where deptno = 30
    -> into outfile '/var/lib/mysql-files/emp.dmp'
    -> fields enclosed by '"' terminated by '|' escaped by '\\'
    -> lines terminated by '\n';
Query OK, 3 rows affected (0.00 sec)
```

以上各个子句中，fields 子句中的 enclosed by 表示括住列值的符号，terminated by 表示列值之间的分隔符，escaped by 表示空值的引导符，用两个"\"，目的是去除"\"的转义功能，lines 子句中的 terminated by 表示使用换行符作为行的结束标记。

查看以上命令创建的导出文件。

```
[root /var/lib/mysql-files 2021-03-03 21:00:46]
# cat emp.dmp
"ALLEN"|"1981-02-02"|"1600.00"|"300.00"
"WARD"|"1981-02-22"|"1250.00"|"500.00"
"MARTIN"|"1981-09-28"|"1250.00"|"1400.00"
"BLAKE"|"1981-05-01"|"2850.00"|\N
"TURNER"|"1981-09-08"|"1500.00"|"0.00"
"JAMES"|"1981-12-03"|"950.00"|\N
```

执行导入操作之前，与导出行的结构相同的表要预先存在。为此，先创建空表 t，由上述导出操作中的列构成。

```
mysql> create table t as
    -> select ename, hiredate, sal, comm from emp where 1 = 0;
```

再执行 load data infile 命令，导入 emp.dmp 的数据至 t 表。注意，最后两行的各选项取值要与导出时的相同。

```
mysql> load data infile '/var/lib/mysql-files/emp.dmp'
    -> into table t
    -> fields enclosed by '"' terminated by '|' escaped by '\\'
    -> lines terminated by '\n';
```

查询验证 t 表的导入数据。

```
mysql> select * from t;
```

19.4.3 使用 mysqlimport 导入

使用 mysqlimport 导入以上 emp.dmp 文件。
先在 db1 数据库中创建 emp 表。

```
mysql> create table db1.emp
    -> (ename char(20), hiredate date, sal numeric(7, 2), comm numeric(7,2));
```

使用 mysqlimport 执行以下命令，把 emp.dmp 中的数据导入 db1.emp，注意分隔符的指定方式与 load data infile 的区别。

```
# mysqlimport db1 --fields-terminated-by='|' --fields-enclosed-by='"' \
> --fields-escaped-by='\' /var/lib/mysql-files/emp.dmp
```

19.4.4 使用 mysqldump 导出

mysqldump 导出数据行，需要附加 --tab 选项，--tab 只能指定为系统参数 secure_file_priv 的值，secure_file_priv 默认为 /var/lib/mysql-files/。使用此方式，每个表导出为两个文本文件，即创建空表的 .sql 文件，以及由数据行构成的 .txt 文件，文件名称与表名相同。

```
# mysqldump db emp --tab=/var/lib/mysql-files/ \
> --fields-terminated-by='|' --fields-enclosed-by='"' --fields-escaped-by='\'
```

以上操作导出的文件如下。

```
# ls
emp.sql  emp.txt
```

19.5 mysqlpump 执行逻辑备份

mysqldump 和 mysqlpump 的主要功能是把 MySQL 实例中的内容（数据库或表）导出为 SQL 命令，重建相关数据库或表时，只要在 mysql 客户端中执行这些 SQL 命令即可。mysqlpump 和 mysqldump 的用法相似，但 mysqlpump 效率更高，这里主要介绍 mysqlpump 的用法。

导出和恢复的命令如下：

mysqlpump [*options*] > dump.sql

mysql [*options*] < dump.sql

19.5.1 mysqlpump 执行导出

除非特别说明，本节的语法形式说明和所有示例也适用于 mysqldump 工具。

mysqlpump 的常用方式有以下几种。

- mysqlpump [other_opts] *db_name table1 table2* ⋯*tablen*：导出一个数据库的多个表；
- mysqlpump [other_opts] *db_name*：导出数据库内的所有表；
- mysqlpump [other_opts] --databases *db1 db2* ⋯ *dbn*：导出多个数据库；
- mysqlpump [other_opts] --all-databases：导出所有用户数据库及 mysql 系统数据库。

导出内容默认显示在屏幕上，为了以后恢复数据，可以使用输出重定向将其导出为 SQL 脚本文件，如导出一个数据库为文件：

mysqlpump [*other_opts*] *db_name* > *file_name*

mysqlpump 工具的导出内容包括建库命令、建表命令及对表添加数据行的 insert 命令。mysqldump 工具只在使用 --databases 和 --all-databases 选项时，才包括建库命令。

导出 db 数据库的 dept 表，导出内容显示在屏幕上。

```
# mysqlpump db dept
```

下面几个示例指定不同选项导出不同内容，并在当前目录下创建 SQL 脚本文件。

导出 db 数据库的 emp 及 dept 表。

```
# mysqlpump db emp dept > emp_dept.sql
```

导出 db 和 db1 数据库。

```
# mysqlpump --databases db db1 > db_db1.sql
```

导出所有数据库。

```
# mysqlpump --all-databases > all.sql
```

完成导出和导入的完整过程。

导出 db 数据库。

```
# mysqlpump db > db.sql
Enter password:
Dump progress: 1/3 tables, 0/21 rows
Dump completed in 133
```

删除 db 数据库模拟其丢失。

```
mysql> drop database db;
Query OK, 3 rows affected (0.03 sec)
mysql> exit
Bye
```

然后执行输入重定向命令将其恢复。

```
# mysql < db.sql
```

也可以在连接服务器后,在 mysql 中执行以下命令(执行之前,先删除 db 数据库):

```
mysql> source db.sql
```

19.5.2 mysqlpump 的常用选项

除了 19.5.1 节介绍的选项外,本节再介绍几个 mysqlpump 常用选项。

- --users:导出创建用户的命令。

下面示例为只导出创建用户的命令,附加 --exclude-databases=% 表示不导出任何数据库。

```
# mysqlpump --exclude-databases=% --users > user.sql
```

mysqldump 也支持此选项。

- --default-parallelism:指定线程并行度,默认为 2。

把并行度设置为 5。

```
# mysqlpump --all-databases --default-parallelism=5 > all_db.sql
```

此选项不适用于 mysqldump。

- --include-tables/--exclude-tables:指定导出内容包括/不包括的表。
- --include-databases/--exclude-databases:指定导出内容包括/不包括的数据库。

以上选项 mysqldump 都不支持。

表名和数据库名称之间以逗号隔开,可以使用"%"和"_"在名称中作为通配符。

导出所有用户数据库,即使用 --all-databases 选项时,不包括 mysql 系统数据库。

```
# mysqlpump --all-databases --exclude-databases=mysql > all_user_db.sql
```

mysqldump 的 --ignore-table 选项与以上的 --exclude-tables 选项相似,指定导出操作略过的表或视图,需要包含数据库名称及表名,如 --ignore-table=db.emp,若要略过多个表,则需使用此参数多次。

- --no-create-db:导出内容不包括建库命令。
- --no-create-info:导出内容不包括建表命令。
- --skip-dump-rows:不导出表的行,即只导出建库建表命令。

只导出对 emp 表添加行的命令。

```
# mysqlpump db emp --no-create-db --no-create-info > emp_rows.sql
```
以上选项中,mysqldump不支持--skip-dump-rows,可以使用--no-data完成相同功能。

19.6 使用mysqlbinlog导出二进制日志文件

mysqlbinlog把二进制日志文件(binary log)转换为SQL语句。
数据库出现故障时,完成恢复的过程一般分两步。
第1步:恢复mysqldump执行的全库备份;
第2步:把从备份开始到数据库出现故障期间执行的操作重新执行一遍。
经过以上两步,即可把数据库还原至出现故障的时刻,保证数据不丢失。
对数据库的操作都保存在二进制日志文件中,完成上面第2步的操作,需要使用mysqlbinlog把二进制日志文件中的相关重做数据转换为SQL语句,然后在mysql中执行。
为了考查二进制日志文件中的内容,先切换至新的日志文件。

```
mysql> flush binary logs;
```
执行show master status查看二进制日志文件的当前状态。

```
mysql> show master status;
+---------------+----------+
| File          | Position |
+---------------+----------+
| binlog.000087 |      156 |
+---------------+----------+
1 row in set (0.00 sec)
```

由以上结果可知,当前正在使用87号日志文件,新的重做数据由文件中的第156字节处开始写入。
连续执行下面几个SQL命令。

```
mysql> update db.emp set sal = 8000 where empno = 7369;
mysql> delete from db.dept where deptno = 40;
mysql> create table db.t(a int, b char(15));
```
再次查看二进制日志文件状态。

```
mysql> show master status;
+---------------+----------+
| File          | Position |
+---------------+----------+
| binlog.000087 |     1014 |
+---------------+----------+
1 row in set (0.00 sec)
```

由以上结果可知,二进制文件中的第156字节至第1 013字节对应以上3个SQL命令产生的重做数据。
下面几个示例说明mysqlbinlog工具的具体用法,在执行下面的命令之前,切换至数据目录,即/var/lib/mysql。
导出多个文件内容。
导出86号和87号日志文件。

```
# mysqlbinlog binlog.000086 binlog.000087
```

根据偏移量范围导出内容。

导出 87 号日志文件中由 320 字节处起至 500 字节处止的重做数据。

```
# mysqlbinlog binlog.000087 --start-position=320 --stop-position=500
```

根据时间范围导出。

指定时间范围为 2021-02-26 10:09:32 至 2021-02-26 10:15:33。

```
# mysqlbinlog binlog.000087 --start-datetime='2021-02-26 10:09:32' \
> --stop-datetime='2020-02-26 10:15:33'
```

19.7 基于时点的恢复

基于时点的恢复(point-in-time recovery)把数据库恢复至故障时刻,不丢失数据。其操作过程包括先恢复之前的 mysqldump 全库备份,然后再执行由备份操作开始至故障时刻期间产生的重做数据。执行重做数据时,需要先用 mysqlbinlog 工具将其转换为 SQL 命令。

下面的实验模拟一个基于时点的完整备份恢复过程。

实验过程以 db 数据库为例,数据库中包括 emp、dept、salgrade 3 个表。

```
mysql> show tables;
+-------------+
| Tables_in_db |
+-------------+
| dept        |
| emp         |
| salgrade    |
+-------------+
3 rows in set (0.00 sec)
```

使用 mysqldump 对 db 数据库执行全库备份,附加--flush-logs 选项,在备份操作开始之前,切换二进制日志文件,使备份开始后产生的重做数据存入新的 92 号日志文件中。

```
# mysqldump --databases db --flush-logs > db.sql
mysql> show master status;
+---------------+----------+
| File          | Position |
+---------------+----------+
| binlog.000092 |      156 |
+---------------+----------+
1 row in set (0.00 sec)
```

在 db 数据库中连续执行以下命令,模拟备份后的数据库操作。

```
mysql> create table db.t(a int, b char(10));
mysql> insert into db.t values(1,'aaa');
mysql> insert into db.t values(2,'bbb');
mysql> insert into db.t values(3,'ccc');
```

切换二进制日志文件,使 92 号日志文件只包含以上操作记录。

```
mysql> flush binary logs;
```

删除 db 数据库,模拟其出现故障。

```
mysql> drop database db;
Query OK, 4 rows affected (0.03 sec)
```

下面执行恢复过程。

执行 db.sql 恢复 db 数据库的初始状态。

```
mysql> source db.sql
```

执行以上操作后,db 数据库已恢复至备份时的状态。

```
mysql> use db
Database changed
mysql> show tables;
+---------------+
| Tables_in_db  |
+---------------+
| dept          |
| emp           |
| salgrade      |
+---------------+
3 rows in set (0.00 sec)
```

执行 92 号日志文件中的重做数据,恢复备份开始后产生的新数据。

```
[root /var/lib/mysql 2021-02-26 12:33:04]
# mysqlbinlog binlog.000092 | mysql db
```

确认在 db 数据库备份中不存在的 t 表及其数据已经被恢复。

```
mysql> show tables;
+---------------+
| Tables_in_db  |
+---------------+
| dept          |
| emp           |
| salgrade      |
| t             |
+---------------+
4 rows in set (0.00 sec)

mysql> select * from t;
+------+------+
| a    | b    |
+------+------+
|    1 | aaa  |
|    2 | bbb  |
|    3 | ccc  |
+------+------+
3 rows in set (0.00 sec)
```

第 20 章 用户和权限管理

一个用户要访问数据库,首先要创建用户账号,依需要设置其属性,然后根据此用户要执行的操作类型对其赋予合适的权限。

本章主要内容包括:
- MySQL 用户的特点;
- 预置用户;
- 用户管理;
- 密码管理;
- 权限管理;
- 用角色简化权限管理;
- 用户及权限信息的查询。

20.1 MySQL 用户的特点

MySQL 的用户都是服务器层面的用户,可以赋予其访问整个服务器的权限,也可以赋予其访问数据库或数据库内对象的权限。

MySQL 中不存在数据库或数据库对象属主的说法。用户创建对象后,这个对象与此用户无关,这个用户也没有操作这个对象的权限。删除用户不影响其创建的对象。

MySQL 的用户没有默认表空间的属性。

MySQL 的用户标识由@符号分隔的两部分构成:user_name@host_name,前一部分为用户名,后一部分为此用户连接服务器时,可登录的客户端机器名。host_name 可以使用通配符"%",指定所有机器或一个子网内的所有机器。

20.2 预 置 用 户

在软件安装后,MySQL 会创建以下 4 个预置用户:root、mysql.sys、mysql.session、mysql.infoschema,其 host_name 都为 localhost,即这几个用户只能在本地登录服务器。

root 为用于执行管理任务的最高权限用户,可以在服务器执行任何操作。安装完毕后,第一次启动 mysqld 服务时,root 用户的口令会自动生成,以此口令连接服务器后,需修改才能执行其他操作。

其他 3 个用户都是系统内部使用的用户,为锁定状态。
- mysql.sys:sys 数据库对象的 DEFINER(创建者)。
- mysql.session:各插件用此用户访问服务器。
- mysql.infoschema:information_schema 数据库对象的 DEFINER。

DEFINER 即数据库创建者的名称。

用户信息存储于 mysql 系统数据库的 user 表中,初次安装 MySQL 软件,启动 mysqld 服务后,可以执行以下查询得到全部预置用户。

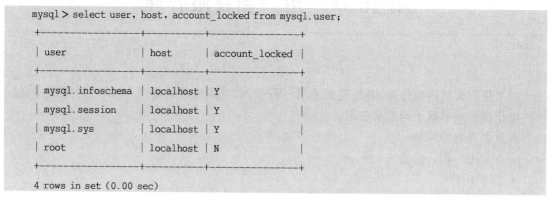

20.3 一个关于用户及权限的简单示例

一个人要访问数据库,要对这个人创建对应的用户,并赋予其适当的访问权限。

下面用一个简单的示例演示创建一个新用户,并使其可以查询 db 数据库的所有表及视图的过程。包含下面 3 个步骤:创建用户,赋予用户权限,以新用户连接数据库进行验证。

以下操作以 root 用户执行。

创建用户 user1,未指定客户端名称,默认为%,即可以从任何客户端登录服务器,密码指定为 User1@2020。

```
mysql> create user user1 identified by 'User1@2020';
Query OK, 0 rows affected (0.00 sec)
```

赋予 user1 用户对 db 数据库所有表及视图的查询权限(所有表及视图以 * 号表示)。

```
mysql> grant select on db.* to user1;
Query OK, 0 rows affected (0.01 sec)
```

以 user1 用户连接服务器,然后切换至 db 数据库,查询 db 数据库的 dept 表(使用-s 选项,使 mysql 不显示提示信息,此时查询结果也不显示网格)。

```
# mysql -u user1 -p'User1@2020' -s
mysql: [Warning] Using a password on the command line interface can be insecure.
mysql> use db
mysql> select * from dept;
deptno  dname       loc
10      ACCOUNTING  NEW YORK
20      RESEARCH    DALLAS
30      SALES       CHICAGO
40      OPERATIONS  BOSTON
```

20.4 用户管理

用户管理包括创建用户,删除用户,设置和改变用户的密码,设置密码复杂度、重用规则、有效时长,设置资源使用策略等任务。

20.4.1 MySQL 的用户名标识

MySQL 的用户名称 user_name@host_name 由两部分构成,user_name 为用户名,host_name 为此用户登录服务器使用的客户端机器名或 IP 地址。

host_name 可以使用通配符"％"表示所有机器或子网内的所有机器,如 u1@192.168.199.135 表示 u1 用户只能在 IP 地址为 192.168.199.135 的客户端登录服务器,u1@％表示 u1 用户可以在任何客户端登录服务器,而 u1@192.168.199.％表示 u1 用户可以使用 192.168.199 子网内的任何客户端登录服务器。

若用户名标识省略 host_name,则默认为"％",即对使用的客户端不做限制,这与其他数据库产品的做法一致。对于用户名标识,除非特殊情形,本书都省略 host_name。

20.4.2 创建和删除用户

使用 create user 命令创建用户,其中的 identified by 子句用于设置密码。

创建用户 user1,不指定客户端名称,默认为"％",即可以从任何客户端登录服务器,密码指定为 User1@2020。

```
mysql> create user user1 identified by 'User1@2020';
Query OK, 0 rows affected (0.00 sec)
```

若删除用户,可以执行 drop user 命令。

```
mysql> drop user user1;
Query OK, 0 rows affected (0.01 sec)
```

默认情况下,MySQL 要求用户密码长度不小于 8 个字符,需要包含大小写字母、数字,以及特殊符号。

默认情况下,密码永不过期,修改密码时,不限制重用旧密码,也不需要输入原密码。

若以上默认设置不满足安全要求,管理员可以在创建用户时,设置其密码的有效时长和旧密码重用的规则;也可以通过设置系统参数,在服务器范围改变以上默认规则。

20.5 密码管理

密码管理包括设置加密算法,过期、重用规则,以及试错次数上限与锁定天数等内容。

下面先演示创建用户时,如何设置各种密码属性,然后说明与密码管理相关的系统参数。

20.5.1 密码加密算法

MySQL 8.0 支持 3 种密码加密算法:caching_sha2_password,sha256_password,以及 mysql_native_password,3 种算法都以插件形式安装。

系统参数 default_authentication_plugin 指定服务器使用的密码验证插件,其值默认为 caching_sha2_password。

mysql_native_password 使用的是 MySQL 引入插件之前的密码散列算法,重新以插件形式实现。

caching_sha2_password 和 sha256_password 都以 SHA-256 算法为基础,只是前者比后

者效率更高,应用也更广泛。sha256_password 算法已经过时,未来版本会删除。

创建用户时,可以指定其密码加密算法。

如下面命令,在创建用户时,设置 user1 的密码加密算法为 mysql_native_password。

```
mysql> create user user1 identified with mysql_native_password by 'User1@2020';
Query OK, 0 rows affected (0.02 sec)
```

如果需要使用第三方工具操作 MySQL,如 Navicat,有可能因其不是最新版本,从而不支持 MySQL 8.0 的 caching_sha2_password 算法,如果不方便更新版本,在创建用户时,可以使用 mysql_native_password 算法,第三方工具一般都会支持此算法。

注意,使用 AppStream 安装的 MySQL 服务器,默认使用 mysql_native_password 插件。

20.5.2 设置密码过期和重用规则

密码重用规则包括重用旧密码的限制,以及修改密码之前,是否需要输入原密码。

限制重用旧密码使用两种方式,一是指定天数,二是指定旧密码个数。

密码过期后,用户登录服务器,执行操作时会收到以下错误信息:

```
mysql> use db
ERROR 1820 (HY000): You must reset your password using ALTER USER statement before executing this statement.
```

必须重置密码才能执行操作。

下面的实验过程都以创建 user1 用户为例,执行每个命令前,请先删除 user1 用户。

创建 user1 用户,指定密码在 90 天后过期。

```
mysql> create user user1 identified by 'User1@2020' password expire interval 90 day;
```

指定密码永不过期。

```
mysql> create user user1 identified by 'User1@2020' password expire never;
```

指定重置密码时,不能使用最近 180 天以内使用过的密码。

```
mysql> create user user1 identified by 'User1@2020' password reuse interval 180 day;
```

指定不能重用最近使用过的 5 个密码。

```
mysql> create user user1 identified by 'User1@2020' password history 5;
```

修改密码时,要求输入原密码,MySQL 8.0.13 开始支持此功能。

```
mysql> create user user1 identified by 'User1@2020' password require current;
```

user1 修改自己的密码时,使用 replace 子句输入之前的旧密码。

```
mysql> alter user user1 identified by 'User1@2021' replace 'User1@2020';
```

20.5.3 设置密码试错次数上限及锁定天数

为了防止非法用户无休止地试探密码,可以设置密码试错次数的上限,以及超过试错次数上限后的锁定天数。

创建 user1 用户,指定其密码试错次数上限为 3,若 3 次输入均为错误密码,则将其锁定 3 天。

```
mysql> create user user1 identified by 'User1@2020'
    -> failed_login_attempts 3 password_lock_time 3;
```

管理员可以执行下面的命令解锁用户。

```
mysql> alter user user1 account unlock;
```

20.5.4 查询用户的密码相关属性

先创建一个用户,同时对其设置密码属性。

```
mysql> create user user1 identified with mysql_native_password
    -> by 'User1@2020'
    -> password expire interval 90 day
    -> password reuse interval 180 day
    -> password require current
    -> failed_login_attempts 3 password_lock_time 3;
```

用户的密码属性存储在 mysql 系统库的 user 表中,密码试错次数和锁定天数存储在 JSON 列 User_attributes 中。

```
mysql> select
    -> user, host, plugin, password_expired, password_lifetime, account_locked,
    -> Password_reuse_history, Password_reuse_time,
    -> Password_require_current, User_attributes
    -> from mysql.user
    -> where user = 'user1' and host = '%'
    -> \G
*************************** 1. row ***************************
                    user: user1
                    host: %
                  plugin: mysql_native_password
        password_expired: N
       password_lifetime: 90
          account_locked: N
  Password_reuse_history: NULL
     Password_reuse_time: 180
Password_require_current: Y
         User_attributes: {"Password_locking": {"failed_login_attempts": 3, "password_lock_time_days": 3}}
1 row in set (0.00 sec)
```

以上 Password_reuse_history 为 NULL,表示对重用旧密码不做限制。

20.5.5 设置密码相关系统参数

要使所有用户在密码限制方面满足某些基本要求,可以通过设置相关系统参数解决。

设置密码有效天数为 90 天。

```
mysql> set persist default_password_lifetime = 90;
```

设置旧密码可重用天数下限为 60 天。

```
mysql> set persist password_reuse_interval = 60;
```

设置旧密码可重用次数下限为 5。

```
mysql> set persist password_history = 5;
```

以上各参数默认值均为 0,即对相关属性不做限制。

设置用户修改自己的密码时,需要输入原密码。

```
mysql> set persist password_require_current = on;
```

password_require_current 的默认值为 off。

20.5.6 设置密码复杂度规则

MySQL 8.0 引入了控件技术(component),每个控件视作一个服务,mysqld 也可以看作一个控件。MySQL 8.0 使用 validate_password 控件实现密码复杂度检查。使用 MySQL 官方 yum 源安装的 MySQL 服务器,validate_password 控件默认已安装激活,使用 AppStream 安装的 MySQL 服务器,默认未安装 validate_password 控件,可以执行以下命令安装。

```
mysql> install component 'file://component_validate_password';
```

卸载控件,只需把 install 改为 uninstall。

validate_password 控件安装后,会产生以下 7 个系统参数用来设置密码复杂度规则。

```
mysql> show variables like 'validate_password%';
+--------------------------------------+--------+
| Variable_name                        | Value  |
+--------------------------------------+--------+
| validate_password.check_user_name    | ON     |
| validate_password.dictionary_file    |        |
| validate_password.length             | 8      |
| validate_password.mixed_case_count   | 1      |
| validate_password.number_count       | 1      |
| validate_password.policy             | MEDIUM |
| validate_password.special_char_count | 1      |
+--------------------------------------+--------+
7 rows in set (0.00 sec)
```

(1) validate_password.check_user_name

用于设置是否检查密码包含了用户名,可选值为 on 和 off,默认为 on。此参数独立于 validate_password.policy。

(2) validate_password.dictionary_file

用于设置是否检查密码包含了字典文件中的词汇,可选值为空或字典文件路径。默认为空,即不检查。若为相对路径,则默认以数据目录开始。此参数若非空,则只在 validate_password.policy 为 2(或 strong)时生效,此时 MySQL 会禁止密码的所有子串使用字典文件内的字符串。字典文件为/usr/share/mysql-8.0/dictionary.txt。

(3) validate_password.length

设置密码的最小长度。其值不能小于以下表达式的值:

number_count+special_char_count+2 * mixed_case_count

若以上表达式中的 3 个参数均为默认值 1,则 validate_password.length 至少要设置为 4。

(4) validate_password.mixed_case_count

设置密码中必须包含的大小写字母个数。大写字母和小写字母个数都不能低于此值。默认为 1,即大写字母和小写字母至少各 1 个。

(5) validate_password.number_count

设置密码中数字的最小个数。

(6) validate_password.policy

用于设置密码复杂度级别。可选值为 0、1、2（或 low、medium、strong）。

0 或 low，只要求密码长度不小于 validate_password.length 的值；1 或 medium，在满足 low 级别的基础上，还要满足对数字、大小写字母以及特殊字符的最小个数要求；2 或 strong，在满足 medium 级别的基础上，若 validate_password.dictionary_file 参数值非空，还要满足密码的子串不能出现在字典文件中的要求。

(7) validate_password.special_char_count

设置密码中特殊字符的最小个数。

MySQL 的密码复杂度规则是针对服务器全局的，不能对个别用户设置单独的密码复杂度规则。

20.6 对用户设置资源限制

以下 4 个选项对用户设置资源限制：max_queries_per_hour、max_updates_per_hour、max_connections_per_hour、max_user_connections。前 2 个选项分别设置 1 个小时内可以执行的 select、update 操作最大次数，第 3 个选项用于设置每小时连接的最大次数，第 4 个选项用于设置一个用户同时连接的最大个数。

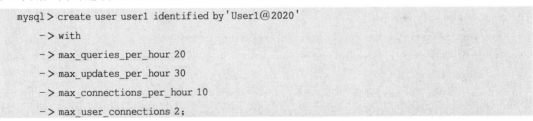

查询以上 user1 的各种资源属性，注意这里的列名与设置用户的资源属性选项不同。

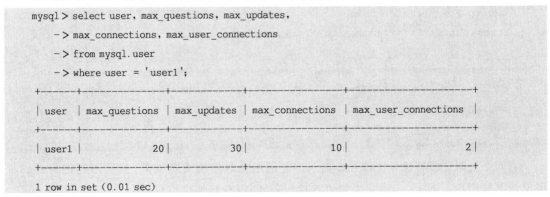

20.7 修改用户属性

修改用户属性的命令与创建用户指定其属性相似，只要把 create 改为 alter。

```
mysql> alter user user1 identified with caching_sha2_password by 'User1@2021'
    -> with max_user_connections 10
    -> password expire interval 100 day
    -> failed_login_attempts 5 password_lock_time 3;
```

在语法顺序方面,注意资源限制选项要放在密码管理属性之前。

20.8 权限管理

一个用户在数据库能执行的操作由其被赋予的权限决定。本节介绍 MySQL 权限管理方面的知识。MySQL 的权限分为若干层次(即生效范围),权限以操作种类命名。赋予用户权限时,只需指定权限名称和生效层次。

20.8.1 MySQL 中的权限层次和名称

MySQl 的权限按其生效范围,分为以下几个层次。
- global:有效范围为服务器全局。
- database:有效范围为单个数据库。
- table 或 routine:有效范围为单个表或存储过程、函数。
- column:有效范围为表的指定列。

MySQL 支持的所有权限名称可以通过以下查询得到。

```
mysql> show privileges\G
*************************** 1. row ***************************
Privilege: Alter
  Context: Tables
  Comment: To alter the table
*************************** 2. row ***************************
Privilege: Alter routine
  Context: Functions,Procedures
  Comment: To alter or drop stored functions/procedures
...
*************************** 58. row ***************************
Privilege: REPLICATION_APPLIER
  Context: Server Admin
  Comment:
58 rows in set (0.00 sec)
```

有两个权限比较特殊:all 和 usage。all 表示相关对象的所有权限,只是不能把得到的权限赋予其他用户。usage 表示除了可以连接服务器,没有其他操作权限。

20.8.2 赋予和撤销权限

赋予用户权限使用 grant 命令,撤销用户权限使用 revoke 命令,命令语法如下:

grant *privs* on *db_name.tbl_name* to *user_name*
revoke *privs* on *db_name.tbl_name* from *user_name*

privs 为权限名称；*user_name* 为用户名；关键字 on 之后为操作对象，也表示权限的有效范围；*db_name* 和 *tbl_name* 分别表示数据库名称和表名称，两者都可以用 * 表示所有数据库或所有表。

若 *db_name* 和 *tbl_name* 都为 *，即操作对象为 *.*，则表示全局权限。赋予全局权限后，重新登录才生效。

若 *tbl_name* 为 *，即操作对象为 *db._name.*，则表示数据库权限。

若两者都为确定值，则表示对象级权限，即操作对象为表或存储过程、函数。

下面给出几个赋权实例。

对 user1 用户赋予全局 all 权限。

```
mysql> grant all on *.* to user1;
```

以上命令若附加 with grant option 子句，则被赋予权限的用户，可以把得到的权限再赋予其他用户，这样，user1 就与 root 用户的权限几乎相同了（user1 还缺少一个权限 PROXY ON "@"，这里的"@"表示所有 user、所有 host，即 root 用户可以代理其他任意用户）。

```
mysql> grant all on *.* to user1 with grant option;
```

撤销 user1 的所有全局权限。

```
mysql> revoke all on *.* from user1;
```

以上命令其实也会撤销其他级别的权限，即所有权限都会撤销，而不限于全局权限，执行后，用户的权限只剩下 usage，即只有连接至服务器的权限。

赋予 user1 用户对 db 数据库中的表 emp 的所有操作权限。

```
mysql> grant all on db.emp to user1;
```

赋予 user1 用户在 db 数据库创建表的权限。

```
mysql> grant create on db.* to user1;
```

以上命令也赋予了 user1 用户创建数据库 db 的权限，即若建库，则名称是固定的。

若表名也指定，如以下命令，则只赋予 user1 用户创建 t 表的权限，即表名是固定的。

```
mysql> grant create on db.t to user1;
```

赋予 user1 用户对 emp 表的 insert、select、update 权限。

```
mysql> grant insert, select, update on db.emp to user1;
```

对于列级权限，需要把列名写在操作名称后面。

```
mysql> grant select(ename, sal) on db.emp to user1;
```

对于存储过程和函数，其执行权限为 execute，下面的命令把 big_table 的执行权限赋予 user1。

```
mysql> grant execute on procedure db.big_table to user1;
```

注意在赋予存储过程执行权限时，不要遗漏 procedure 关键字。

除了 all on *.*，只有用 grant 命令赋予的权限，才能用 revoke 命令撤销。

以下命令赋予 user1 的 select 全局权限，然后撤销 db 数据库的查询权限。

```
mysql> grant select on *.* to user1;
Query OK, 0 rows affected (0.00 sec)

mysql> show grants for user1;
```

```
+----------------------------------------+
| Grants for user1@ %                    |
+----------------------------------------+
| GRANT SELECT ON *.* TO `user1`@`%`     |
+----------------------------------------+
1 row in set (0.00 sec)
```

以下两个命令都不被支持。

```
mysql> revoke select on db.* from user1;
ERROR 1141 (42000): There is no such grant defined for user 'user1' on host '%'
mysql> revoke all on db.* from user1;
ERROR 1141 (42000): There is no such grant defined for user 'user1' on host '%'
```

虽然之前并未赋予 all on *.*，但撤销 all on *.* 是支持的。

```
mysql> revoke all on *.* from user1;
Query OK, 0 rows affected (0.00 sec)
```

执行上面的命令后，user1 的各种权限都会被撤销。

20.8.3 全局权限的部分撤销

对用户赋予全局权限时，其生效范围是服务器上所有数据库的所有对象。若要 user1 的全局权限对几个数据库例外，如 user1 的 all 全局权限不包括 mysql 系统库，下面命令，MySQL 并不支持。

```
mysql> grant all on *.* to user1;
Query OK, 0 rows affected (0.01 sec)

mysql> revoke all on mysql.* from user1;
ERROR 1141 (42000): There is no such grant defined for user 'user1' on host '%'
```

为了解决以上问题，可以开启 MySQL 8.0.16 新引入的系统参数 partial_revokes，其默认值为 off，即不支持对权限部分撤销。将其设置为 on 后，则支持部分撤销权限。

```
mysql> set persist partial_revokes = on;
Query OK, 0 rows affected (0.00 sec)


mysql> revoke all on mysql.* from user1;
Query OK, 0 rows affected (0.00 sec)
```

以上命令只撤销了 mysql 系统数据库的操作权限，保留了其他数据库的操作权限。

下面的示例对其他全局权限执行部分撤销。

撤销所有权限，重新开始实验过程。

```
mysql> revoke all on *.* from user1;
Query OK, 0 rows affected (0.00 sec)


mysql> show grants for user1;
+----------------------------------------+
| Grants for user1@ %                    |
+----------------------------------------+
```

```
| GRANT USAGE ON *.* TO `user1`@`%` |
+-----------------------------------+
1 row in set (0.00 sec)
```

赋予 user1 用户 4 个全局权限：select、insert、update、delete。

```
mysql> grant select, insert, update, delete on *.* to user1;
Query OK, 0 rows affected (0.00 sec)

mysql> show grants for user1;
+------------------------------------------------------------------+
| Grants for user1@%                                               |
+------------------------------------------------------------------+
| GRANT SELECT, INSERT, UPDATE, DELETE ON *.* TO `user1`@`%`       |
+------------------------------------------------------------------+
1 row in set (0.00 sec)
```

撤销 user1 用户对 mysql 系统库的 select 和 update 权限，以及 db 数据库的 select 权限。

```
mysql> revoke select, update on mysql.* from user1;
Query OK, 0 rows affected (0.01 sec)

mysql> revoke select on db.* from user1;
Query OK, 0 rows affected (0.00 sec)
```

查询 user1 的权限情况，可以看到以上部分撤销的权限以 revoke 语句显示在结果中。

```
mysql> show grants for user1;
+------------------------------------------------------------------+
| Grants for user1@%                                               |
+------------------------------------------------------------------+
| GRANT SELECT, INSERT, UPDATE, DELETE ON *.* TO `user1`@`%`       |
| REVOKE SELECT ON `db`.* FROM `user1`@`%`                         |
| REVOKE SELECT, UPDATE ON `mysql`.* FROM `user1`@`%`              |
+------------------------------------------------------------------+
3 rows in set (0.00 sec)
```

在系统表 mysql.user 中，被部分撤销的全局权限保存在 User_attributes 列中。

使用 mysql.user 查看 user1 的全局权限情况。

```
mysql> select * from mysql.user where user = 'user1'\G
*************************** 1. row ***************************
                  Host: %
                  User: user1
             Select_priv: Y
                   ...
Password_require_current: NULL
         User_attributes: {"Restrictions": [{"Database": "db", "Privileges": ["SELECT"]},
{"Database": "mysql", "Privileges": ["SELECT", "UPDATE"]}]}
1 row in set (0.00 sec)
```

此功能仅针对全局权限有效，不支持从数据库级别权限中撤销指定表的操作权限。

20.8.4 查询用户的权限信息

查询用户的权限信息,最方便的方法是利用 show grants 命令。

root 用户可以执行 show grants for *user_name* 查询指定用户的权限。普通用户可以执行 show grants 查询自己的权限。

下面过程以 root 用户执行,先赋予 user1 用户若干权限,然后执行查询验证。

```
mysql> show grants for user1;
+-----------------------------------+
| Grants for user1@%                |
+-----------------------------------+
| GRANT USAGE ON *.* TO `user1`@`%` |
+-----------------------------------+
1 row in set (0.00 sec)

mysql> grant select, update on db.emp to user1;
Query OK, 0 rows affected (0.01 sec)

mysql> grant all on db1.* to user1;
Query OK, 0 rows affected (0.00 sec)

mysql> grant insert on db.emp to user1;
Query OK, 0 rows affected (0.00 sec)

mysql> show grants for user1;
+-----------------------------------------------------------+
| Grants for user1@%                                        |
+-----------------------------------------------------------+
| GRANT USAGE ON *.* TO `user1`@`%`                         |
| GRANT ALL PRIVILEGES ON `db1`.* TO `user1`@`%`            |
| GRANT SELECT, INSERT, UPDATE ON `db`.`emp` TO `user1`@`%` |
+-----------------------------------------------------------+
3 rows in set (0.00 sec)
```

查询结果中,同一对象的权限会组织在一个 grant 语句中。

user1 登录后,可以执行 show grants 命令查询自身的权限。

```
mysql> show grants;
+-----------------------------------------------------------+
| Grants for user1@%                                        |
+-----------------------------------------------------------+
| GRANT USAGE ON *.* TO `user1`@`%`                         |
| GRANT ALL PRIVILEGES ON `db1`.* TO `user1`@`%`            |
| GRANT SELECT, INSERT, UPDATE ON `db`.`emp` TO `user1`@`%` |
+-----------------------------------------------------------+
3 rows in set (0.00 sec)
```

用户及权限信息存储于 mysql 库中的几个系统表：user、db、tables_priv、columns_priv、procs_priv。这几个表称为 grant tables。

user：存储用户属性及全局权限信息。不管是否赋予了权限，每个用户在 mysql.user 表中都对应一行。每个权限一列，以 Y 和 N 标识是否赋予了相关权限。

只有用户被赋予相关级别的权限后，其权限信息才会出现在下面几个表中。

db：存储用户的数据库范围的权限。

tables_priv：存储用户的表级权限。

columns_priv：存储用户的列级权限。

procs_priv：存储用户对存储过程或函数的操作权限。

可以查询以上系统表得到用户的相关权限信息。

查询 user1 用户的全局权限。

先撤销 user1 的所有权限。

```
mysql> revoke all on *.* from user1;
Query OK, 0 rows affected (0.00 sec)

mysql> show grants for user1;
+---------------------------------------+
| Grants for user1@%                    |
+---------------------------------------+
| GRANT USAGE ON *.* TO `user1`@`%`     |
+---------------------------------------+
1 row in set (0.00 sec)
```

对 user1 赋予全局 select 权限、db 数据库的 delete 权限、db.emp 表的 insert 权限、db.emp 表的 sal 列的 update 权限、db.big_table 存储过程的执行权限。

```
mysql> grant select on *.* to user1;
Query OK, 0 rows affected (0.00 sec)

mysql> grant delete on db.* to user1;
Query OK, 0 rows affected (0.00 sec)

mysql> grant insert on db.emp to user1;
Query OK, 0 rows affected (0.01 sec)

mysql> grant update(sal) on db.emp to user1;
Query OK, 0 rows affected (0.03 sec)

mysql> grant execute on procedure db.big_table to user1;
Query OK, 0 rows affected (0.03 sec)
```

执行 show grants 命令查看 user1 的权限情况。

```
mysql> show grants for user1;
+------------------------------------------------------------------+
| Grants for user1@%                                               |
+------------------------------------------------------------------+
| GRANT SELECT ON *.* TO `user1`@`%`                               |
| GRANT DELETE ON `db`.* TO `user1`@`%`                            |
| GRANT INSERT, UPDATE (`sal`) ON `db`.`emp` TO `user1`@`%`        |
| GRANT EXECUTE ON PROCEDURE `db`.`big_table` TO `user1`@`%`       |
+------------------------------------------------------------------+
```

分别查询 mysql 的权限相关系统表。

查询 user 表，得到 user1 的全局权限。

```
mysql> select * from mysql.user where user = 'user1'\G
*************************** 1. row ***************************
                    Host: %
                    User: user1
              Select_priv: Y
              Insert_priv: N
                    ...
Password_require_current: NULL
          User_attributes: NULL
1 row in set (0.00 sec)
```

查询 db 表，得到 user1 的数据库权限。

```
mysql> select * from mysql.db where user = 'user1' and db = 'db'\G
*************************** 1. row ***************************
           Host: %
             Db: db
           User: user1
    Select_priv: N
    Insert_priv: N
    Update_priv: N
    Delete_priv: Y
    Create_priv: N
                ...
1 row in set (0.00 sec)
```

查询 tables_priv，得到 user1 的表级权限。

```
mysql> select * from mysql.tables_priv where user = 'user1'\G
*************************** 1. row ***************************
       Host: %
         Db: db
       User: user1
 Table_name: emp
    Grantor: root@localhost
  Timestamp: 0000-00-00 00:00:00
```

```
    Table_priv: Insert
   Column_priv: Update
1 row in set (0.00 sec)
```

查询 columns_priv,得到 user1 的列级权限。

```
mysql> select * from mysql.columns_priv where user = 'user1';
+------+----+-------+------------+-------------+---------------------+-------------+
| Host | Db | User  | Table_name | Column_name | Timestamp           | Column_priv |
+------+----+-------+------------+-------------+---------------------+-------------+
| %    | db | user1 | emp        | sal         | 0000-00-00 00:00:00 | Update      |
+------+----+-------+------------+-------------+---------------------+-------------+
1 row in set (0.00 sec)
```

查询 mysql.procs_priv,得到 user1 的存储过程执行权限。

```
mysql> select * from mysql.procs_priv where user = 'user1'\G
*************************** 1. row ***************************
        Host: %
          Db: db
        User: user1
Routine_name: big_table
Routine_type: PROCEDURE
     Grantor: root@localhost
    Proc_priv: Execute
   Timestamp: 0000-00-00 00:00:00
1 row in set (0.00 sec)
```

information_schema 系统库的 4 个视图也可以用来查询权限信息,包括 user_privileges、schema_privileges、table_privileges、column_privileges,分别查询用户的全局权限、数据库权限、表级权限和列级权限。information_schema 数据库中没有存储过程权限相关的视图。

下面几个命令分别查询 user1 的各种权限。

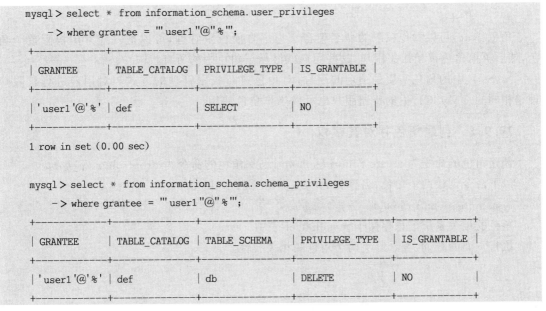

```
1 row in set (0.00 sec)

mysql> select * from information_schema.table_privileges
    -> where grantee = "'user1'@'%'"
    -> \G
*************************** 1. row ***************************
       GRANTEE: 'user1'@'%'
 TABLE_CATALOG: def
  TABLE_SCHEMA: db
    TABLE_NAME: emp
PRIVILEGE_TYPE: INSERT
  IS_GRANTABLE: NO
1 row in set (0.00 sec)

mysql> select * from information_schema.column_privileges
    -> where grantee = "'user1'@'%'"
    -> \G
*************************** 1. row ***************************
       GRANTEE: 'user1'@'%'
 TABLE_CATALOG: def
  TABLE_SCHEMA: db
    TABLE_NAME: emp
   COLUMN_NAME: sal
PRIVILEGE_TYPE: UPDATE
  IS_GRANTABLE: NO
1 row in set (0.00 sec)
```

20.9 角 色

角色(role)可以看作权限的命名集合。为了方便权限管理,可以把一些常用权限授予角色,然后再把角色授予相关用户,这些用户就继承了角色中的所有权限。

MySQL 中的角色实质是锁定的用户,对角色进行权限管理的方式与对用户进行权限管理是相同的。MySQL 中的普通用户也可以当作角色使用。

20.9.1 创建角色并对其赋权

创建角色的命令为 create role *role_name*;删除角色的命令为 drop role *role_name*。

以下命令创建两个角色,名称为 rw_db_emp 和 rw_db_dept。

```
mysql> create role rw_db_emp, rw_db_dept;
```

角色的相关系统信息存储于 mysql.user 表中。

执行下面的查询,查看以上两个角色的相关属性。

```
mysql> select host, user, authentication_string, account_locked
    -> from mysql.user
    -> where user like 'rw%';
+------+-----------+-----------------------+----------------+
| host | user      | authentication_string | account_locked |
+------+-----------+-----------------------+----------------+
| %    | rw_db_dept|                       | Y              |
| %    | rw_db_emp |                       | Y              |
+------+-----------+-----------------------+----------------+
2 rows in set (0.00 sec)
```

可以发现,角色实质上是密码为空的锁定用户。

对两个角色分别赋予读写 db 数据库的 emp 表和 dept 表的权限。

```
mysql> grant select, update on db.emp to rw_db_emp;
Query OK, 0 rows affected (0.00 sec)

mysql> grant select, update on db.dept to rw_db_dept;
Query OK, 0 rows affected (0.01 sec)
```

可以像对用户一样,查询角色的权限。

```
mysql> show grants for rw_db_emp;
+-------------------------------------------------------------+
| Grants for rw_db_emp@%                                      |
+-------------------------------------------------------------+
| GRANT USAGE ON *.* TO `rw_db_emp`@`%`                       |
| GRANT SELECT, UPDATE ON `db`.`emp` TO `rw_db_emp`@`%`       |
+-------------------------------------------------------------+
2 rows in set (0.00 sec)

mysql> show grants for rw_db_dept;
+-------------------------------------------------------------+
| Grants for rw_db_dept@%                                     |
+-------------------------------------------------------------+
| GRANT USAGE ON *.* TO `rw_db_dept`@`%`                      |
| GRANT SELECT, UPDATE ON `db`.`dept` TO `rw_db_dept`@`%`     |
+-------------------------------------------------------------+
2 rows in set (0.00 sec)
```

20.9.2 角色的赋予及撤销

对用户赋予和撤销角色的命令分别为:

grant *role_name* to *user_name*

revoke *role_name* from *user_name*

其中 *role_name* 和 *user_name* 分别为角色名称和用户名称。

下面的实验只说明角色的赋予,角色的撤销比较简单,请读者自行实验。

把 user1 的权限都撤销,然后把 rw_db_emp 和 rw_db_dept 两个角色赋予 user1。

```
mysql> revoke all on *.* from user1;
Query OK, 0 rows affected (0.01 sec)

mysql> grant rw_db_emp, rw_db_dept to user1;
Query OK, 0 rows affected (0.00 sec)
```

用户被赋予的角色信息存储在 mysql.role_edges 系统表中。

```
mysql> select * from mysql.role_edges;
+-----------+------------+---------+---------+-------------------+
| FROM_HOST | FROM_USER  | TO_HOST | TO_USER | WITH_ADMIN_OPTION |
+-----------+------------+---------+---------+-------------------+
| %         | rw_db_dept | %       | user1   | N                 |
| %         | rw_db_emp  | %       | user1   | N                 |
+-----------+------------+---------+---------+-------------------+
2 rows in set (0.00 sec)
```

其中的 from_user 为角色名称，to_user 为用户名称。

也可以执行 show grants 命令，查看 user1 的角色信息。

```
mysql> show grants for user1;
+-------------------------------------------------------------------+
| Grants for user1@%                                                |
+-------------------------------------------------------------------+
| GRANT USAGE ON *.* TO `user1`@`%`                                 |
| GRANT `rw_db_dept`@`%`,`rw_db_emp`@`%` TO `user1`@`%`             |
+-------------------------------------------------------------------+
2 rows in set (0.00 sec)
```

20.9.3 激活角色

继续 20.9.2 节实验过程。

虽然 user1 已被赋予了两个角色，若此用户查询 emp，可以发现并不具备角色中的权限。

```
# mysql -u user1 -p -s
Enter password:
mysql> select ename, sal from db.emp where ename = 'SMITH';
ERROR 1142 (42000): SELECT command denied to user 'user1'@'localhost' for table 'emp'
```

出现以上问题的原因是 user1 的角色尚未激活。用户的角色激活后，才会生效。

current_role() 函数返回当前用户激活的角色。以 user1 用户执行下面的查询，可以确认，当前用户被赋予的角色都未激活。

```
mysql> select current_role();
+----------------+
| current_role() |
+----------------+
| NONE           |
+----------------+
1 row in set (0.00 sec)
```

以 user1 执行 set role 命令激活角色 rw_db_emp,再次查看激活的角色。

```
mysql> set role rw_db_emp;
Query OK, 0 rows affected (0.00 sec)

mysql> select current_role();
+----------------+
| current_role() |
+----------------+
| `rw_db_emp`@`%` |
+----------------+
1 row in set (0.00 sec)
```

查询 emp 表,这时不再报错。

```
mysql> select ename, sal from db.emp where ename = 'SMITH';
+-------+--------+
| ename | sal    |
+-------+--------+
| SMITH | 800.00 |
+-------+--------+
1 row in set (0.04 sec)
```

执行 show grants 查看其权限信息。

```
mysql> show grants;
+-------------------------------------------------------------------+
| Grants for user1@%                                                |
+-------------------------------------------------------------------+
| GRANT USAGE ON *.* TO `user1`@`%`                                 |
| GRANT SELECT, UPDATE ON `db`.`emp` TO `user1`@`%`                 |
| GRANT `rw_db_dept`@`%`,`rw_db_emp`@`%` TO `user1`@`%`             |
+-------------------------------------------------------------------+
3 rows in set (0.00 sec)
```

由以上结果可以发现,rw_db_emp 中的权限已经生效。

如果除一个之外,如 rw_db_emp,激活其他所有角色,可以执行下面的命令。

```
mysql> set role all except rw_db_emp;
```

如果关闭所有角色的激活状态,可以执行下面的命令。

```
mysql> set role none;
```

以上设置只对当前连接生效,退出连接或服务器重启后会失效。

如果需要角色在用户登录服务器时立刻生效,可以设置以下系统参数。

```
mysql> set persist activate_all_roles_on_login = on;
```

20.9.4 用户的默认角色

用户的默认角色在登录时即激活,不需要另外设置。使用 set default role 命令设置默认角色。以下命令把 rw_db_emp 角色设置为 user1 的默认角色。

```
mysql> set default role rw_db_emp to user1;
```

以下命令把 user1 被赋予的所有角色都设置为默认。

```
mysql> set default role all to user1;
```

以下命令使得 user1 没有默认角色。

```
mysql> set default role none to user1;
```

以上设置会存储于 mysql.default_roles 中,在服务器重启之后,依然生效。

20.9.5 设置公共角色

公共角色即所有用户默认具备的角色,一般把最基本的查询权限赋予公共角色。需要注意的是,公共角色也需要激活才生效,另外,不能撤销用户的公共角色。

MySQL 使用系统参数 mandatory_roles 指定公共角色。

把 rw_db_emp 和 rw_db_dept 两个角色设置为公共角色。

```
mysql> set persist mandatory_roles ='rw_db_emp, rw_db_dept';
```

20.9.6 角色相关信息查询

管理员可以像对用户一样使用 show grants 命令查询角色中的权限,或使用 mysql 系统数据库中的相关系统表,查询角色中的权限。

root 用户使用 show grants 命令,查询 rw_db_emp 的权限信息。

```
mysql> show grants for rw_db_emp;
+-----------------------------------------------------------+
| Grants for rw_db_emp@%                                    |
+-----------------------------------------------------------+
| GRANT USAGE ON *.* TO `rw_db_emp`@`%`                     |
| GRANT SELECT, UPDATE ON `db`.`emp` TO `rw_db_emp`@`%`     |
+-----------------------------------------------------------+
2 rows in set (0.00 sec)
```

root 用户使用 mysql.tables_priv,查询 rw_db_emp 的表级权限信息。

```
mysql> select * from mysql.tables_priv
    -> where user = 'rw_db_emp'
    -> \G
*************************** 1. row ***************************
       Host: %
         Db: db
       User: rw_db_emp
 Table_name: emp
    Grantor: root@localhost
  Timestamp: 0000-00-00 00:00:00
 Table_priv: Select,Update
Column_priv:
1 row in set (0.00 sec)
```

root 用户查询 user1 具备的角色信息,可以使用 show grants 命令。

```
mysql> show grants for user1;
+-----------------------------------------------------------+
| Grants for user1@%                                        |
+-----------------------------------------------------------+
| GRANT USAGE ON *.* TO `user1`@`%`                         |
| GRANT SELECT, UPDATE ON `db`.`dept` TO `user1`@`%`        |
| GRANT SELECT, UPDATE ON `db`.`emp` TO `user1`@`%`         |
| GRANT `rw_db_dept`@`%`,`rw_db_emp`@`%` TO `user1`@`%`     |
+-----------------------------------------------------------+
4 rows in set (0.00 sec)
```

上面查询结果的最后一行即用户被赋予的角色信息。

也可以使用 mysql.role_edges 系统表。

```
mysql> select * from mysql.role_edges;
+-----------+------------+---------+---------+-------------------+
| FROM_HOST | FROM_USER  | TO_HOST | TO_USER | WITH_ADMIN_OPTION |
+-----------+------------+---------+---------+-------------------+
| %         | rw_db_dept | %       | user1   | N                 |
| %         | rw_db_emp  | %       | user1   | N                 |
| %         | rw_db_emp  | %       | user2   | N                 |
+-----------+------------+---------+---------+-------------------+
3 rows in set (0.00 sec)
```

其中的 from_user 为角色名称，to_user 为用户名称。

root 用户查询 user1 用户的默认角色信息，可以使用 mysql.default_roles。

```
mysql> select * from mysql.default_roles;
+------+-------+-------------------+-------------------+
| HOST | USER  | DEFAULT_ROLE_HOST | DEFAULT_ROLE_USER |
+------+-------+-------------------+-------------------+
| %    | user1 | %                 | rw_db_emp         |
+------+-------+-------------------+-------------------+
1 row in set (0.00 sec)
```

user1 用户查询自身已激活角色的信息，可以使用 current_role() 函数。

```
mysql> select current_role();
+------------------------------------+
| current_role()                     |
+------------------------------------+
| `rw_db_dept`@`%`,`rw_db_emp`@`%`   |
+------------------------------------+
1 row in set (0.01 sec)
```

参 考 文 献

[1] Jesper Wisborg Krogh. MySQL 8 Query Performance Tuning[M]. New York：Apress，2020.

[2] 小孩子4919. MySQL是怎样运行的——从根儿上理解MySQL[M]. 北京：人民邮电出版社，2020.

[3] Daniel Bartholomew. MariaDB and MySQL Common Table Expressions and Window Functions Revealed[M]. New York：Apress，2017.

[4] Guy Harrison，Steven Feuerstein. MySQL Stored Procedure Programming[M]. Sebastopol，CA：O'Reilly Media，Inc.，2006.